THE
UNIVERSE AROUND US

BY
SIR JAMES JEANS, M.A., D.S.C., LL.D., F.R.S.

NEW YORK: THE MACMILLAN COMPANY
CAMBRIDGE, ENGLAND: AT THE UNIVERSITY PRESS

COPYRIGHT, 1929,
By THE MACMILLAN COMPANY.

All rights reserved, including the right of reproduction
in whole or in part in any form.

Set up and electrotyped.
Published September, 1929.
Reprinted November, 1929.
December, 1929. Twice.

SET UP BY BROWN BROTHERS LINOTYPERS
PRINTED IN THE UNITED STATES OF AMERICA
BY THE FERRIS PRINTING COMPANY

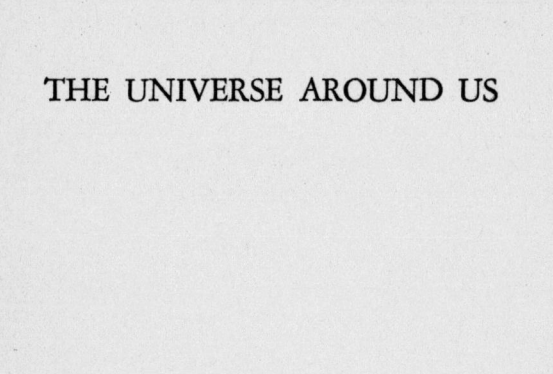

THE UNIVERSE AROUND US

PREFACE

THE present book contains a brief account, written in simple language, of the methods and results of modern astronomical research, both observational and theoretical. Special attention has been given to problems of cosmogony and evolution, and to the general structure of the universe. My ideal, perhaps never wholly attainable, has been that of making the entire book intelligible to readers with no special scientific knowledge.

Parts of the book cover the same ground as various lectures I have recently delivered to University and other audiences, including a course of wireless talks I gave last autumn. It has been found necessary to rewrite these almost in their entirety, so that very few sentences remain in their original form, but those who have asked me to publish my lectures and wireless talks will find the substance of them in the present book.

J. H. JEANS.

Dorking,
May 1, 1929

CONTENTS

		PAGE
INTRODUCTION—THE STUDY OF ASTRONOMY		1

CHAPTER		
I.	EXPLORING THE SKY	16
II.	EXPLORING THE ATOM	86
III.	EXPLORING IN TIME	141
IV.	CARVING OUT THE UNIVERSE	184
V.	STARS	238
VI.	BEGINNINGS AND ENDINGS	305
	INDEX	333

PLATES

PLATE		FACING PAGE
I.	The Milky Way in the neighbourhood of the Southern Cross	23
II.	Planetary Nebulae: 1. N.G.C. 2022 2. N.G.C. 6720 (The Ring Nebula) 3. N.G.C. 1501 4. N.G.C. 7662	28
III.	The Nebula in Cygnus	29
IV.	The Great Nebula in Andromeda (*M* 31)	30
V.	A Nebula in Andromeda (N.G.C. 891) seen edge-on	31
VI.	The "Horse's Head" south of ζ Orionis, in the Great Nebula in Orion	36
VII.	The Trifid Nebula in Sagittarius	42
VIII.	Stellar Spectra	48
IX.	The Globular Cluster in Hercules (*M* 13)	60
X.	The Region of ϱ Ophiuchi	62
XI.	Magnification of part of the outer regions of the Great Nebula in Andromeda (*M* 31)	66
XII.	Magnification of the central region of the Great Nebula in Andromeda (*M* 31)	67
XIII.	The tracks of α- and β-particles	106
XIV.	Collisions of α-particles with atoms	111
XV.	The Nebula N.G.C. 4594 in Virgo The Nebula N.G.C. 7217	194

PLATES

PLATE		FACING PAGE
XVI.	A sequence of Nebular Configurations: 1. N.G.C. 3379 2. N.G.C. 4621 3. N.G.C. 3115 4. N.G.C. 4594 in Virgo 5. N.G.C. 4565 in Berenice's hair	196
XVII.	The "Whirlpool" Nebula (M 51) in Canes Venatici	198
XVIII.	The Nebula M 81 in Ursa Major	199
XIX.	The Nebula M 101 in Ursa Major	200
XX.	The Nebula M 33 in Triangulum	201
XXI.	The Lesser and Greater Magellanic Clouds	204
XXII.	Two Nebulae (N.G.C. 4395, 4401) suggestive of tidal action	226
XXIII.	The twin Nebulae N.G.C. 4567-8 The Nebula N.G.C. 7479	227
XXIV.	Saturn and its System of Rings	234

THE UNIVERSE AROUND US

INTRODUCTION
The Study of Astronomy

ON the evening of January 7, 1610, a fateful day for the human race, Galileo Galilei, Professor of Mathematics in the University of Padua, sat in front of a telescope he had made with his own hands.

More than three centuries previously, Roger Bacon, the inventor of spectacles, had explained how a telescope could be constructed so as "to make the stars appear as near as we please." He had shewn how a lens could be so shaped that it would collect all the rays of light falling on it from a distant object, bend them until they met in a focus and then pass them on through the pupil of the eye on to the retina. Such an instrument would increase the power of the human eye, just as an ear trumpet increases the power of the human ear by collecting all the waves of sound which fall on a large aperture, bending them, and passing them through the orifice of the ear on to the eardrum.

Yet it was not until 1608 that the first telescope had been constructed by Lippershey, a Flemish spectacle-maker. On hearing of this instrument Galileo had set to work to discover the principles of its construction and had soon made himself a telescope far better than the original. His instrument had created no small sensation in Italy. Such extraordinary stories had been told of its powers that he had been commanded to take it to Venice and exhibit it to the Doge and Senate. The citizens of Venice had then seen the most aged of their Senators climbing the highest bell-towers to spy through the telescope at ships which were too far out at sea to be seen at all without its help. The telescope admitted about a hundred times as much light as the unaided human eye, and, according to Galileo, it shewed an object at fifty miles as clearly as if it were only five miles away.

The absorbing interest of his new instrument had almost driven from Galileo's mind a problem to which he had at one time given much thought. Over two thousand years previously, Pythagoras and Philolaus had taught that the earth is not fixed in space but rotates on its axis every twenty-four hours, thus causing the alternation of day and night. Aristarchus of Samos, perhaps the greatest of all the Greek mathematicians, had further maintained that the earth not only turned on its axis, but also described a yearly journey round the sun, this being the cause of the cycle of the seasons.

Then these doctrines had fallen into disfavour. Aristotle had pronounced against them, asserting that the earth formed a fixed centre to the universe. Later Ptolemy had explained the tracks of the planets across the sky in terms of a complicated system of cycles and epicycles; the planets moved in circular paths around moving points which themselves moved in circles around an immoveable earth. The Church had given its sanction and active support to these doctrines. Indeed, it is difficult to see what else it could have done, for it seemed almost impious to suppose that the great drama of man's fall and redemption, in which the Son of God had Himself taken part, could have been enacted on any lesser stage than the very centre of the Universe.

INTRODUCTION—THE STUDY OF ASTRONOMY

Yet, even in the Church, the doctrine had not gained universal acceptance. Oresme, Bishop of Lisieux, and Cardinal Nicholas of Cusa, had both declared against it, the latter writing in 1440:

> I have long considered that this earth is not fixed, but moves as do the other stars. To my mind the earth turns upon its axis once every day and night.

At a later date those who held these views incurred the active hostility of the Church, and in 1600 Giordano Bruno was burned at the stake. He had written:

> It has seemed to me unworthy of the divine goodness and power to create a finite world, when able to produce beside it another and others infinite; so that I have declared that there are endless particular worlds similar to this of the earth; with Pythagoras I regard it as a star, and similar to it are the moon, the planets and other stars, which are infinite in number, and all these bodies are worlds.

The most weighty attack on orthodox doctrine had, however, been delivered neither by theologians nor philosophers, but by the Polish astronomer, Nicolaus Copernicus (1473-1543). In his great work *De revolutionibus orbium coelestium* Copernicus had shewn that Ptolemy's elaborate structure of cycles and epicycles was unnecessary, because the tracks of the planets across the sky could be explained quite simply by supposing that the earth and the planets all moved round a fixed central sun. The sixty-six years which had elapsed since this book was published had seen these theories hotly debated, but they were still neither proved nor disproved.

Galileo had already found that his new telescope provided a means of testing astronomical theories. As soon as he had turned it on to the Milky Way, a whole crowd of legends and fables as to its nature and structure had vanished into thin air; it proved to be nothing more than a swarm of faint stars scattered like golden dust on the black background of the sky. Another glance through the telescope had disclosed the true nature of the moon. It had on it mountains which cast shadows, and so proved, as Giordano Bruno had maintained, to be a world like our own. What if the telescope should now in some way prove able to decide between the orthodox doctrine that the earth formed the hub of the universe, and the new doctrine that the earth was only one of a number of bodies, all circling round the sun like moths round a candle-flame?

And now Galileo catches Jupiter in the field of his telescope and sees four small bodies circling around the great mass of the planet—like moths round a candle-flame. What he sees is an exact replica of the solar system as imagined by Copernicus, and it provides direct visual proof that such systems are at least not alien to the architectural plan of the universe. And yet, strangely enough, he hardly sees the full implications of his discovery at once; he merely avers that he has discovered four new planets which chase one another round and round the known planet Jupiter.

Final and complete understanding comes nine months later when he observes the phases of Venus. Venus might have been self-luminous, in which case she would always appear as a full circle of light. If she were not self-luminous but moved in a Ptolemaic epicycle, then, as Ptolemy had himself pointed out, she could never shew more than half her surface illuminated. On the other hand, the Copernican view of the solar system required that both Venus and Mercury should exhibit "phases" like those of the moon, their shining surfaces ranging in appearance from crescent-shape

INTRODUCTION—THE STUDY OF ASTRONOMY

through half moon to full moon, and then back through half moon to crescent shape. That such phases were not shewn by Venus had indeed been urged as an objection to the Copernican theory.

Galileo's telescope now shews that, as Copernicus had foretold, Venus passes through the full cycle of phases, so that, in Galileo's own words, we "are now supplied with a determination most conclusive, and appealing to the evidence of our senses," that "Venus, and Mercury also, revolve around the sun, as do also all the rest of the planets, a truth believed indeed by the Pythagorean school, by Copernicus, and by Kepler, but never proved by the evidence of our senses, as is now proved in the case of Venus and Mercury."

These discoveries of Galileo made it clear that Aristotle, Ptolemy and the majority of those who had thought about these things in the last 2000 years had been utterly and hopelessly wrong. In estimating his position in the universe, man had up to now been guided mainly by his own desires, and his self-esteem; long fed on boundless hopes, he had spurned the simpler fare offered by patient scientific thought. Inexorable facts now dethroned him from his self-arrogated station at the centre of the universe; henceforth he must reconcile himself to the humble position of the inhabitant of a speck of dust, and adjust his views on the meaning of human life accordingly.

The adjustment was not made at once. Human vanity, reinforced by the authority of the Church, contrived to make a rough road for those who dared draw attention to the earth's insignificant position in the universe. Galileo was forced to abjure his beliefs. Well on into the eighteenth century the ancient University of Paris taught that the motion of the earth round the sun was a convenient *but false* hypothesis, while the newer American Universities of Harvard and Yale taught the Ptolemaic and Copernician systems of astronomy side by side as though they were equally tenable. Yet men could not keep their heads buried in the sand for ever, and when at last its full implications were accepted, the revolution of thought initiated by Galileo's observations of January 7, 1610, proved to be the most catastrophic in the history of the race. The cataclysm was not confined to the realms of abstract thought; henceforth human existence itself was to appear in a new light, and human aims and aspirations would be judged from a different standpoint.

This oft-told story has been told once again, in the hope that it may serve to explain some of the interest taken in astronomy to-day. The more mundane sciences prove their worth by adding to the amenities and pleasures of life, or by alleviating pain or distress, but it may well be asked what reward astronomy has to offer. Why does the astronomer devote arduous nights, and still more arduous days, to studying the structure, motions and changes of bodies so remote that they can have no conceivable influence on human life?

In part at least the answer would seem to be that many have begun to suspect that the astronomy of to-day, like that of Galileo, may have something to say on the enthralling question of the relation of human life to the universe in which it is placed, and on the beginnings, meaning and destiny of the human race. Bede records how, some twelve centuries ago, human life was compared in poetic simile to the flight of a bird through a warm hall in which men sit feasting, while the winter storms rage without.

INTRODUCTION—THE STUDY OF ASTRONOMY

The bird is safe from the tempest for a brief moment, but immediately passes from winter to winter again. So man's life appears for a little while, but of what is to follow, or of what went before, we know nothing. If, therefore, a new doctrine tells us something certain, it seems to deserve to be followed.

These words, originally spoken in advocacy of the Christian religion, describe what is perhaps the main interest of astronomy to-day. Man

> only knowing
> Life's little lantern between dark and dark

wishes to probe further into the past and future than his brief span of life permits. He wishes to see the universe as it existed before man was, as it will be after the last man has passed again into the darkness from which he came. The wish does not originate solely in mere intellectual curiosity, in the desire to see over the next range of mountains, the desire to attain a summit commanding a wide view, even if it be only of a promised land which he may never hope himself to enter; it has deeper roots and a more personal interest. Before he can understand himself, man must first understand the universe from which all his sense perceptions are drawn. He wishes to explore the universe, both in space and time, because he himself forms part of it, and it forms part of him.

We may well admit that science cannot at present hope to say anything final on the questions of human existence and human destiny, but this is no justification for not becoming acquainted with the best that it has to offer. It is rare indeed for science to give a final "Yes" or "No" answer to any question propounded to her. When we are able to put a question in such a definite form that either of these answers could be given in reply, we are generally already in a position to supply the answer ourselves. Science advances rather by providing a succession of approximations to the truth, each more accurate than the last, but each capable of endless degrees of higher accuracy. To the question, "where does man stand in the universe?" the first attempt at an answer, at any rate in recent times, was provided by the astronomy of Ptolemy: "at the centre." Galileo's telescope provided the next, and incomparably better, approximation: "man's home in space is only one of a number of small bodies revolving round a huge central sun." Nineteenth-century astronomy swung the pendulum still further in the same direction, saying: "there are millions of stars in the sky, each similar to our sun, each doubtless surrounded, like our sun, by a family of planets on which life may be kept in being by the light and heat received from its sun." Twentieth-century astronomy suggests, as we shall see, that the nineteenth century had swung the pendulum too far; life now seems to be more of a rarity than our fathers thought, or would have thought if they had given free play to their intellects.

We are setting out to explain the approximation to the truth provided by twentieth-century astronomy. No doubt it is not the final truth, but it is a step on towards it, and unless we are greatly in error it is very much nearer to the truth than was the teaching of nineteenth-century astronomy. It claims to be nearer the truth, not because the twentieth-century astronomer claims to be better at guessing than his predecessors of the nineteenth century, but because he has incomparably more facts at his disposal. Guessing has gone out of fashion in science; it was at best a poor substitute for knowledge, and modern science, eschewing guessing severely, confines itself, except on very rare occasions, to

INTRODUCTION—THE STUDY OF ASTRONOMY

ascertained facts and the inferences which, so far as can be seen, follow unequivocally from them.

It would of course be futile to pretend that the whole interest of astronomy centres round the questions just mentioned. Astronomy offers at least three other groups of interest which may be described as utilitarian, scientific and aesthetic.

At first astronomy, like other sciences, was studied for mainly utilitarian reasons. It provided measures of time, and enabled mankind to keep a tally on the flight of the seasons; it taught him to find his way across the trackless desert, and later, across the trackless ocean. In the guise of astrology, it held out hopes of telling him his future. There was nothing intrinsically absurd in this, for even to-day the astronomer is largely occupied with foretelling the future movements of the heavenly bodies, although not of human affairs—a considerable part of the present book will consist of an attempt to foretell the future, and predict the final end, of the material universe. Where the astrologers went wrong was in supposing that terrestrial empires, kings and individuals formed such important items in the scheme of the universe that the motions of the heavenly bodies could be intimately bound up with their fates. As soon as man began to realise, even faintly, his own insignificance in the universe, astrology died a natural and inevitable death.

The utilitarian aspect of astronomy has by now shrunk to very modest proportions. The national observatories still broadcast the time of day, and help to guide ships across the ocean, but the centre of astronomical interest has shifted so completely that the remotest of nebulae arouse incomparably more enthusiasm than "clock-stars," and the average astronomer totally neglects our nearest neighbours in space, the planets, for stars so distant that their light takes hundreds, thousands, or even millions, of years to reach us.

Recently, astronomy has acquired a new scientific interest through establishing its position as an integral part of the general body of science. The various sciences can no longer be treated as distinct; scientific discovery advances along a continuous front which extends unbroken from electrons of a fraction of a millionth of a millionth of an inch in diameter, to nebulae whose diameters are measured in hundreds of thousands of millions of millions of miles. A gain of astronomical knowledge may add to our knowledge of physics and chemistry, and *vice versa*. The stars have long ago ceased to be treated as mere points of light. Each is now regarded as an experiment on a heroic scale, a high temperature crucible in which nature herself operates with ranges of temperature and pressure far beyond those available in our laboratories, and permits us to watch the results. In so doing, we may happen upon properties of matter which have eluded the terrestrial physicist, owing to the small range of physical conditions at his command. For instance matter exists in nebulae with a density at least a million times lower than anything we can approach on earth, and in certain stars at a density nearly a million times greater. How can we expect to understand the whole nature of matter from laboratory experiments in which we can command only one part in a million million of the whole range of density known to nature?

Yet for each one who feels the purely scientific appeal of astronomy, there are probably a dozen who are attracted by its aesthetic appeal. Many even of those who seek after knowledge for its own sake, driven by that intellectual curi-

INTRODUCTION—THE STUDY OF ASTRONOMY

osity which provides the fundamental distinction between themselves and the beasts, find their main interest in astronomy, as the most poetical and the most aesthetically gratifying of the sciences. They want to exercise their faculties and imaginations on something remote from everyday trivialities, to find an occasional respite from "the long littleness of life," and they satisfy their desires in contemplating the serene immensities of the outer universe. To many, astronomy provides something of the vision without which the people perish.

Before proceeding to describe the results of the modern astronomer's survey of the sky, let us try to envisage in its proper perspective the platform from which his observations are made.

Later on, we shall see how the earth was born out of the sun, something like two thousand millions of years ago. It was born in a form in which we should find it hard to recognise the solid earth of to-day with its seas and rivers, its rich vegetation and overflowing life. Our home in space came into being as a globe of intensely hot gas on which no life of any kind could either gain or retain a foothold.

Gradually this globe of gas cools down, becoming first liquid, then plastic. Finally its outer crust solidifies, rocks and mountains forming a permanent record of the irregularities of its earlier plastic form. Vapours condense into liquids, and rivers and oceans come into being, while the "permanent" gases form an atmosphere. Gradually the earth assumes a condition suited to the advent of life, which finally appears, we know not how, whence or why.

It is not easy to estimate the time since life first appeared on earth, but it can hardly have been more than a small fraction of the whole 2000 million years of the earth's existence. Still, there was probably life on earth at least 300 million years ago. The first life appears to have been wholly aquatic, but gradually fishes changed into reptiles, reptiles into mammals, and finally man emerged from mammals. The evidence favours a period of about 300,000 years ago for this last event. Thus life has inhabited the earth for only a fraction of its existence, and man for only a tiny fraction of this fraction. To put it in another way, the astronomical time-scale is incomparably longer than the human time-scale—the generations of man, and even the whole of human existence, are only ticks of the astronomer's clock.

Most of the 10,000 or so of generations of men who connect us up with our ape-like ancestry must have lived lives which did not differ greatly from those of their animal predecessors. Hunting, fishing and warfare filled their lives, leaving but little time or opportunity for intellectual contemplation. Then, at last, man began to awake from his long intellectual slumber, and, as civilisation slowly dawned, to feel the need for occupations other than the mere feeding and clothing of his body. He began to discover revelations of infinite beauty in the grace of the human form or the play of light on the myriad-smiling sea, which he tried to perpetuate in carefully chiselled marble or exquisitely chosen words. He began to experiment with metals and herbs, and with the effects of fire and water. He began to notice, and try to understand, the motions of the heavenly bodies, for to those who could read the writing in the sky, the nightly rising and setting of the stars and planets provided evidence that beyond the confines of the earth lay an unknown universe built on a far grander scale.

In this way the arts and sciences came to earth, bringing

INTRODUCTION—THE STUDY OF ASTRONOMY

astronomy with them. We cannot quite say when, but compared even with the age of the human race, they came but yesterday, while in comparison with the whole age of the earth, their age is but a twinkling of the eye.

Scientific astronomy, as distinguished from mere stargazing, can hardly claim an age of more than 3000 years. It is less than this since Pythagoras, Aristarchus and others explained that the earth moved around a fixed sun. Yet the really significant figure for our present purpose is not so much the time since men began to make conjectures about the structure of the universe, as the time since they began to unravel its true structure by the help of ascertained fact. The important length of time is that which has elapsed since that evening in 1610 when Galileo first turned his telescope on to Jupiter—a mere three centuries or so.

We begin to grasp the true significance of these round-number estimates when we re-write them in tabular form. We have:

Age of earth	about 2,000,000,000	years
Age of life on earth	" 300,000,000	"
Age of man on earth	" 300,000	"
Age of astronomical science	" 3,000	"
Age of telescopic astronomy	" 300	"

When the various figures are displayed in this form, we see what a very recent phenomenon astronomy is. Its total age is only a hundredth part of the age of man, only a hundred-thousandth part of the time that life has inhabited the earth. During 99,999 parts out of the 100,000 of its existence, life on earth was hardly concerned about anything beyond the earth. But whereas the past of astronomy is to be measured on the human time-scale, a hundred generations or so of men, there is every reason to expect that its future will be measured on the astronomical time-scale. We shall discuss the probable future stretching before the human race in a later chapter. For the moment it is not unreasonable to suppose that this future will probably be terminated by astronomical causes, so that its length is to be measured on the astronomical time-scale. As the earth has already existed for 2000 million years, it is *à priori* reasonable to suppose that it will exist for at least something of the order of 2000 million years yet to come, and humanity and astronomy with it. Actually we shall find reasons for expecting it to last far longer than this. But if once it is conceded that its future life is to be estimated on the astronomical time-scale, no matter in what exact way, we see that astronomy is still at the very opening of its existence. This is why its message can claim no finality—we are not describing the mature convictions of a man, so much as the first impressions of a newborn babe which is just opening its eyes. Even so they are better than the idle introspective dreamings in which it indulged before it had learned to look around itself and away from itself.

And so we set out to learn what astronomy has to tell us about the universe in which we live our lives. Our inquiry will not be entirely limited to this one science. We shall call upon other sciences, physics, chemistry and geology, as well as the more closely allied sciences of astrophysics and cosmogony, to give help, when they can, in interpreting the message of observational astronomy. The information we shall obtain will be fragmentary. If it must be compared to anything, let it be to the pieces of a jig-saw puzzle. Could we get hold of all the pieces, they would, we are confident, form a single complete consistent picture, but many of them are still missing. It is too much to hope that the incomplete series of pieces we have already found will disclose

INTRODUCTION—THE STUDY OF ASTRONOMY

the whole picture, but we may at least collect them together, arrange them in some sort of methodical order, fit together pieces which are obviously contiguous, and perhaps hazard a guess as to what the finished picture will prove to be when all its pieces have been found and finally fitted together.

CHAPTER I

Exploring the Sky

WE have seen how man, after inhabiting the earth for 300,000 years, has within the last 300 years—the last one-thousandth part of his life on earth—become possessed of an optical means of studying the outer universe. In the present chapter we shall try to describe the impressions he has formed with his newly-awakened eyes. The description will be arranged in a very rough chronological order. This is also an order of increasing telescopic power, or again of seeing further and further into space, so that our order of arrangement might equally be described as one of increasing distance from the sun. We shall not attempt any sort of continuous record, but shall merely mention a few landmarks so as to shew in broad outline the order in which territory was won and consolidated in man's survey of the universe.

THE SOLAR SYSTEM

Our first landmark, or perhaps it is better to say our starting point, is the unravelling of the structure of the solar system by Galileo and his successors.

The sun's family of planets falls naturally into two distinct groups—four "minor planets," Mercury, Venus, the Earth and Mars, which are of small size and near to the sun, and four "major planets," Jupiter, Saturn, Uranus and Neptune, which are large in size and very distant from the sun.

Mercury is nearest of all to the sun; next comes Venus. The orbits of these two planets lie between the earth's orbit and the sun. As seen from the earth, these planets appear to describe relatively small circles round the sun, and so must necessarily appear near to the sun in the sky. As a consequence, they can only be seen either in the early morning, if they happen to rise just before the sun, or in the evening if they set after the sun. The ancients not altogether recognising that the same planets could appear both as morning and evening stars, gave them different names according as they figured as the one or the other. As a morning star Venus was called Phosphoros by the Greeks and Lucifer by the Romans; as an evening star it was called Hesperus by both.

Next beyond the earth, proceeding outward from the sun into space, comes Mars, completing the group of minor planets. Mars, Venus and Mercury are all smaller than the earth in size, although Venus is only slightly so.

There is a wide gap between the orbit of Mars, the outermost of the minor planets, and that of Jupiter, the innermost of the major planets. This is not empty; it is occupied by the orbits of thousands of tiny planets known as asteroids. None of these approaches the earth in size; Ceres, the largest, is only 480 miles in diameter, and only four are known with diameters of more than 100 miles. The planets Mercury, Venus and Mars have all been known from remote antiquity, but the asteroids only entered astronomy with the nineteenth century, Ceres, the first and largest, having been discovered by Piazzi on January 1, 1801.

Beyond the asteroids come the four major planets Jupiter,

Saturn, Uranus and Neptune, all of which are far larger than the earth. Jupiter, the largest, has according to Sampson, a diameter of 88,640 miles, or more than eleven times the diameter of the earth; fourteen hundred bodies of the size of the earth could be packed inside Jupiter, and leave room to spare. Saturn, which comes next in order, is second only to Jupiter in size, having a diameter of about 70,000 miles. These two are by far the largest of the planets. All the others rolled into one would only make a ball a fifth of the size of Saturn, and these together with Saturn would make a ball of only slightly more than half of Jupiter's size.

Although Uranus and Neptune, the two outermost members of the solar system, are far smaller than Jupiter and Saturn, yet each has about four times the diameter of the earth. Jupiter and Saturn form such conspicuous objects in the sky that they have necessarily been known from the earliest times, but Uranus and Neptune are comparatively recent discoveries. Sir William Herschel discovered Uranus quite accidentally in 1781, while looking through his telescope with no motive other than the hope of finding something interesting in the sky. By contrast, Neptune was discovered in 1846 as the result of intricate mathematical calculations, which many at the time regarded as the greatest triumph of the human mind, at any rate since the time of Newton. The honour of its discovery must be apportioned in approximately equal shares between a young Englishman, John Couch Adams, afterwards Professor of Astronomy at Cambridge, and a young French astronomer, Urbain J. J. Leverrier. Both attributed certain vagaries in the observed motion of Uranus to the gravitational pull of an exterior planet, and both set to work to calculate the orbit in which this supposed outer planet must move to explain these vagaries. Both obtained the required orbit, if not with great exactness, at least with sufficient accuracy to indicate whereabouts in the sky the supposed new planet ought to be found.

Adams finished his calculations first, and informed observers at Cambridge as to the part of the sky in which the new planet ought to lie. As a result, Neptune was observed twice, although without being immediately identified as the wanted planet. Before this identification had been established at Cambridge, Leverrier had finished his computations and communicated his results to Galle, an assistant at Berlin, who was able to identify the planet at once, Berlin possessing better star-charts of the region of the sky in question than were accessible at Cambridge.

As far back as 1772, Bode had pointed out a simple numerical relation connecting the distances of the various planets from the sun. This is obtained as follows: Write first the series of numbers

0 1 2 4 8 16 32 64 128

in which each number after the first two is double the preceding. Multiply each by three, thus obtaining

0 3 6 12 24 48 96 192 384

and add four to each, giving

4 7 10 16 28 52 100 196 388

These numbers are very approximately proportional to the actual distances of the planets from the sun, which are (taking the earth's distance to be 10):

3.9	7.2	10.0	15.2	26.5	52.0	95.4	191.9	300.7
Mercury	Venus	Earth	Mars	Asteroids	Jupiter	Saturn	Uranus	Neptune

The law was enunciated before the discovery of Uranus,

[19]

the asteroids, or Neptune, so that it is somewhat remarkable that Uranus and the asteroids, when discovered, fitted fairly accurately into their predicted places. On the other hand the law fails completely for Neptune, and ought strictly to be considered to fail for Mercury also, since the original series of numbers 0, 1, 2, 4, 8, . . . begins in an artificial way. The true mathematical series would of course be $\frac{1}{2}$, 1, 2, 4, 8, . . . each being double the preceding, and this would give 5$\frac{1}{2}$ for the distance of Mercury as against the actual distance of 3.9.

So far no explanation of Bode's law has been given, and it seems more than likely that it is a mere coincidence with no underlying rational explanation.

The outermost planets are at enormous distances from the sun, Neptune being more than 30 times as distant as the Earth. An inhabitant of Neptune, if such existed, would receive only a nine-hundredth part as much light and heat from the sun as an inhabitant of the earth receives.

It can be calculated that, if this were its only source of heat, Neptune's surface would be at the very low temperature indeed of about — 220° Centigrade, but it is possible that it may have internal sources of heat which keep its surface at a higher temperature. The infinitesimal amount of heat which we on earth receive from Jupiter has recently been measured. Its amount shews that the surface of Jupiter is at a temperature of about — 150° Centigrade, which is just about that at which it would be maintained by the sun's heat alone. On the other hand similar measurements assign temperatures of — 150° and — 170° respectively to Saturn and Uranus, both of which are rather higher than would be expected if these planets had no source of heat beyond the sun's radiation. But it seems likely that any sources of internal heat must be quite small, and that all the major planets are very cold indeed. There can be neither seas nor rivers on their surfaces, since all water must be frozen into ice, neither can there be rain or water-vapour in their atmospheres. It has been suggested that the clouds which obscure our view of Jupiter's surface may be condensed particles of carbon-dioxide, or some other gas which boils at temperatures far below the freezing point of water.

The physical conditions of the minor planets are much more like those with which we are familiar on earth. Owing to its greater distance from the sun, Mars is somewhat, but not enormously, colder than the earth. Its day of 24 hours 37 minutes is only slightly longer than our own, so that its surface must experience alternations of warmth by day and cold by night similar to those we find on earth. In the equatorial regions the temperature rises well above the freezing point at noon, probably reaching 50° Fahrenheit or even more. But even here it falls below freezing some time before sunset, and from then until well on in the next day, the climate must be very cold. The polar regions are of course colder still, the temperature of the snowcap which covers the poles being somewhere about — 70° Centigrade or — 94° Fahrenheit—126 degrees of frost!

Venus, being nearer the sun, must have a higher average temperature than the earth. But as each of its days and nights is several weeks of our terrestrial time, the difference between the temperatures of day and night must be far greater than with us, so that its surface must experience great extremes of heat by day and of cold by night. The night temperature appears to be fairly uniformly equal to about — 25° Centigrade or — 13° Fahrenheit. At any point on the planet's surface weeks of this bitterly cold night temperature must alternate with weeks of a roasting day temperature.

Mercury is so near the sun that its average temperature is necessarily far higher than that of the earth. It seems likely that the planet always turns the same face to the sun, just as the moon always turns the same face to the earth. If so the unwarmed half of its surface must be intensely cold, and the warmed half intensely hot. Pettit and Nicholson have measured the amount of heat received on earth from the warmed hemisphere, and find that its temperature must be about 350° Centigrade or 662° Fahrenheit, a temperature which melts lead. The other half of the planet's surface, eternally dark and unwarmed, is probably colder than anything we can imagine.

Galileo's discovery of the four satellites of Jupiter was followed in time by the discovery that every planet was attended by satellites, except the two whose orbits lay inside the earth's. In 1655 Huyghens discovered Titan, the largest of Saturn's satellites, and by 1684 Cassini had discovered four more. Then, after the lapse of a full century, Sir William Herschel discovered two satellites of Uranus in 1787 and two more satellites of Saturn in 1789. We shall discuss the full system of planetary satellites and also the smaller bodies of the solar system—comets, meteors and shooting-stars—in a later chapter, when we come to deal with the way they came into being.

The Galactic System

Our next landmark is the survey of the stars by the two Herschels, Sir William Herschel, the father (1738-1822) and Sir John Herschel, the son (1792-1871). What Galileo had done for the solar system, the two Herschels set out to

PLATE I

Franklin-Adams Chart.

The Milky Way in the neighbourhood of the Southern Cross.

do for the huge family of stars—the "galactic" system, bounded by the Milky Way—of which our sun is a member.

On a clear moonless night the Milky Way is seen to stretch, like a great arch of faint light, from horizon to horizon. It is found to be only part of a full circle of light —the galactic circle—which stretches completely round the earth and divides the sky into two equal halves, forming a sort of celestial "equator," with reference to which astronomers are accustomed to measure latitude and longitude in the sky. Galileo's telescope had shewn that it consists of a crowd of faint stars, each too dim to be seen individually without telescopic aid (see Plate I). And, as might be expected, the proper interpretation of this great belt of faint stars has proved to be fundamental in understanding the architecture of the universe.

If stars were scattered uniformly through infinite space, we should at last come to a star in whatever direction we looked, so that the sky would appear as a uniform blaze of intolerable light. It is true that this would not be the case if light were dimmed or blotted out after travelling a certain distance, but even then, the sky would appear the same in all directions, for there would be no reason why one part of the sky should be more lavishly spangled with stars than another. Thus the existence of the Milky Way shews that the system of the stars does not extend uniformly to infinity. It must have a definite structure, and it was the architecture of this that Sir William Herschel set himself to unravel. The work he did for the northern half of the sky was subsequently extended to the southern hemisphere by his son, Sir John Herschel.

We shall best understand the method employed by the Herschels if we first imagine all the stars in the sky to be intrinsically similar objects. Each would then emit the same amount of light, so that the nearer stars would appear bright, and the further stars faint, merely as an effect of distance. The way in which apparent brightness decreases with distance is of course well known; the law is that of the "inverse square of the distance," which means that the apparent brightness decreases just as rapidly as the square of its distance increases; a star which is twice as distant as a second similar star appears only a quarter as bright, and so on. Thus if all stars emitted the same amount of light, we could estimate the relative distances of any two stars in the sky from their relative brightness. By cutting wires of lengths proportional to the distances of various stars, and pointing these in the directions of the stars to which they referred, we could form a model of the arrangement of the stars in the sky. We should, in fact, know the whole structure of the system of stars except for its scale. To represent the faint stars of the Milky Way, a great number of very long wires would be needed. In the model these would all point towards different parts of the Milky Way, forming a flat wheel-like structure.

The problem which confronted Sir William Herschel was more intricate because he knew that the stars were of different intrinsic brightness as well as at different distances, and both factors combined to produce differences of apparent brightness. One of the main difficulties of astronomy, both to the Herschels and to the astronomer of to-day, is that these two factors have to be disentangled before any definite conclusions are reached.

Herschel found that the number of stars visible in his telescopic-field varied enormously with different directions in space. It was of course greatest when the telescope was pointed at the Milky Way, and fell off, steadily and rapidly, as the telescope was moved away from the Milky Way. Generally speaking, two telescope-fields which were at equal distances from the Milky Way contained about the same number of stars. In the technical language of astronomy, the richness of the star-field depended mainly on the galactic latitude, just as the earth's climate depends mainly on the geographic latitude, and not to any great extent on the longitude.

Fields at different distances from the Milky Way were found to differ in quality as well as in number of stars. The brightest stars of all occurred about equally in all fields, the difference in the fields resulting mainly from faint stars, and particularly the faintest stars of all, becoming enormously more abundant as the Milky Way was approached.

Sir William Herschel rightly interpreted this as shewing that the system of stars surrounding the sun began to thin out within distances reached by his telescope, and that they began to thin out soonest in directions furthest away from the Milky Way. He supposed the general shape of the galactic system of stars to be that of a bun or a biscuit or a watch, the stars being most thickly scattered near the centre, and occurring more sparsely in the outer regions. The plane of the Milky Way of course formed the central plane of the structure. The fact that the Milky Way divides the sky into two almost exactly equal halves suggested to him that the sun must be very nearly in this central plane, and this is confirmed by the recent very refined investigations of Seares and van Rhijn, and others. From the fact that parts of the sky which were equidistant from the Milky Way appeared about equally bright, Herschel inferred that the sun not only lay in the central plane of the system, but was very near to its actual centre. This view has prevailed until quite recently, but the researches of Shapley and Seares now shew it to be untenable.

Fig. 1 shews a cross section of the general kind of structure which Sir William Herschel assigned to the galactic system, although the detailed distribution of stars shewn in the diagram is that given at a much later date (1922) by Kapteyn. It is easy to see how a structure of this type would account for the general appearance of the sky. Those stars

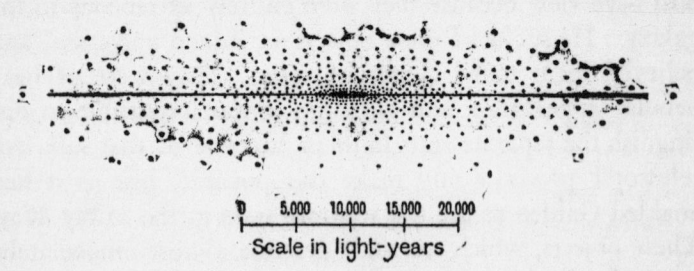

FIG. 1.—The Structure of the Galactic System according to Herschel and Kapteyn.

which appear brightest of all are, generally speaking, the nearest; they are so near that no appreciable thinning out of stars occurs within this distance. For this reason the very bright stars occur in about equal numbers in all directions. The stars which appear very faint are mostly very distant, so distant that the great depth of the system in directions in or near to the galactic plane is brought into play. In such directions, layer after layer of stars, ranged almost endlessly one behind the other, give rise to the apparent concentration of faint stars which we call the Milky Way.

The final acceptance of the Copernican view of the structure of the solar system was in a large measure due to Galileo's discovery of the similar system of Jupiter, which was so situated in space that a terrestrial observer could obtain a bird's-eye view of it as a whole. We can never obtain a bird's-eye view of the solar system as a whole because we can only see it from inside, so that optical proof that such systems could exist, could come only from the discovery of other similar systems, which we could see from outside.

Sir William Herschel believed he had confirmed his own view of the structure of the galactic system in the same way, by discovering similar systems, of which he could obtain a bird's-eye view because they were entirely extraneous to the galaxy. He spoke of these objects as "island universes" and believed them to be clouds of stars. They were of hazy nebular appearance, and although it was impossible to distinguish the separate stars in them, he believed that sufficient telescopic power would make this possible, just as it had enabled Galileo to see the separate stars in the Milky Way. These objects, which we shall describe almost immediately, are generally known as "extra-galactic nebulae" from their position, although we shall frequently find it convenient to use the brief term "great nebulae," to which their immense size fully entitles them.

Nebulae

A telescope exhibits a planet as a disc of appreciable size, and an eye-piece which magnifies 60 times will make Jupiter look as large as the moon. Yet an eye-piece which magnifies 60 times, or any greater number of times, can never make a star look as large as the moon. No magnification within our command causes any star to appear as anything other than a mere point of light. The stars are of course enormously larger than Jupiter, but they are also enormously more distant, and it is the distance that wins.

The telescope nevertheless shews a number of objects which appear bigger than mere points of light. They are generally of a faint, hazy appearance, and so have received the general name of "nebulae." Detailed investigation has shewn that these fall into three distinct classes.

Planetary Nebulae. The first class are generally described as "Planetary Nebulae." There is nothing of a planetary nature about them beyond the fact that, like the planets, they shew as finite discs in a telescope. A few hundreds only of these objects are known, four typical examples being illustrated in Plate II. They lie within the galactic system, and are probably of the nature of stars which have in some way become surrounded by luminous atmospheres of enormous extent. If so, they of course disprove our general statement that no star ever appears as anything but a point of light in a telescope; we must make an exception in favour of the planetary nebulae.

Galactic Nebulae. The second class are generally described as "Galactic Nebulae," examples being shown in Plates III, VI (p. 36) and VII (p. 42). These are completely irregular in shape. Their general appearance is that of huge glowing wisps of gas stretching from star to star, and in effect this is pretty much what they are. Like the planetary nebulae, they lie entirely within the galactic system. Even a cursory glance shews that each irregular nebula contains several stars enmeshed with it; minute telescopic examination often extends the dimensions of the nebula almost indefinitely, so that we may have almost the whole of a constellation wrapped up in a single nebula.

There is but little doubt as to the physical nature of these nebulae. The space between the stars is not utterly void of matter, but is occupied by a thin cloud of gas of a tenuity

PLATE II

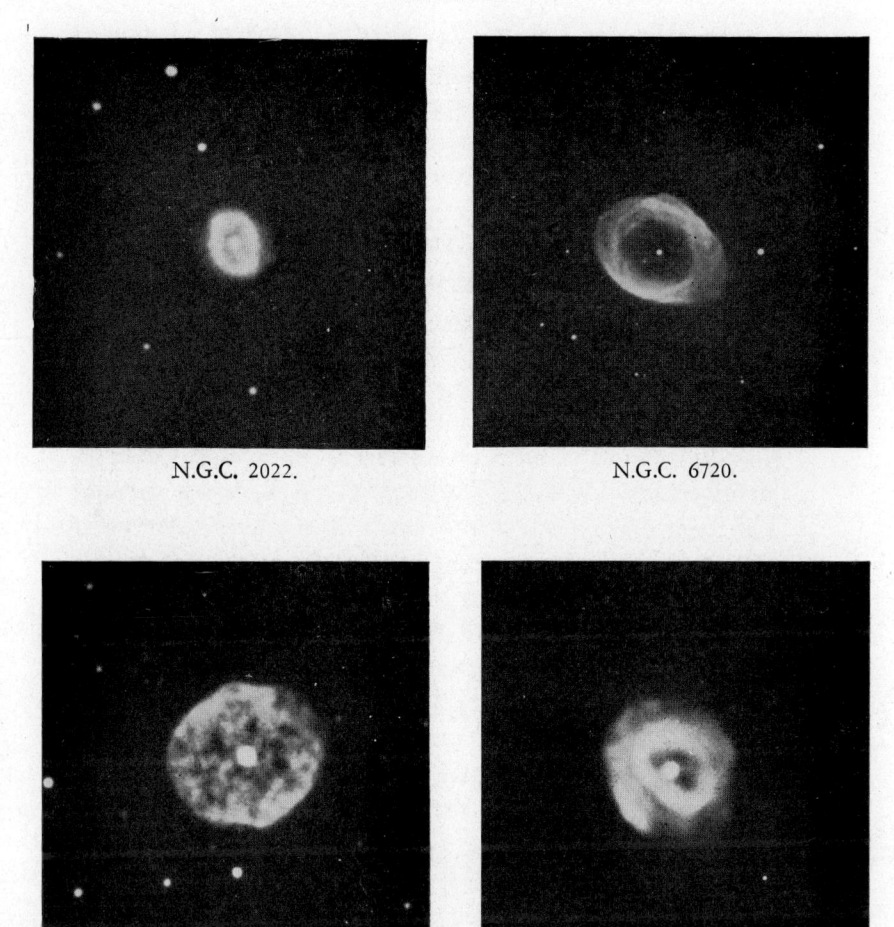

N.G.C. 2022.

N.G.C. 6720.

N.G.C. 1501.

N.G.C. 7662.
Mt. Wilson Observatory.

Planetary Nebulae.

PLATE III

Nebula in Cygnus. *Mt. Wilson Observatory.*

which is generally almost beyond description. Here and there this cloud may be denser than usual; here and there again it may be lighted up and made to incandesce by the radiation of the stars within it. In other places it may be entirely opaque to light, lying like a black curtain across the sky. The variations of density, opacity and luminosity in combination produce all the fantastic shapes and varied degrees of light and shade we see in the galactic nebulae.

This same opacity is responsible for the dark patches which occur in the general arrangement of the stars. A conspicuous example occurs in the part of the Milky Way shewn in Plate I (p. 23). The dark patch, which looks at first like a hole in the system of stars, is graphically described as "The Coal Sack." These black patches in the sky cannot represent actual holes, because it is inconceivable that there should be so many empty tunnels through the stars all pointing exactly earthward, so that we are compelled to interpret them as veils of obscuring matter which dim or extinguish the light of the stars behind them.

Extra-galactic Nebulae. The third class of nebula is of an altogether different nature. Its members are for the most part of definite and regular shape, and shew various other characteristics which make them easy of identification. They used to be called "white nebulae" from the quality of the light they emitted. Later Lord Rosse's giant 6-foot telescope revealed that many of them had a spiral structure; these were called "spiral nebulae." The most conspicuous of all the spiral nebulae is the Great Nebula (M 31) in Andromeda, shewn in Plate IV, which is just, and only just, visible to the naked eye. Marius, observing it telescopically in 1612, described it as looking "like a candle-light seen through horn." Plate V shews a second example, probably of very similar structure, which is viewed from another angle, so as to appear almost exactly edge-on.

It is now abundantly proved that nebulae of this type all lie outside the galactic system, so that the term "extra-galactic nebulae" adequately describes them. Their size is colossal. Either of the photographs shewn in Plates IV and V would have to be enlarged to the size of the whole of Europe before a body of the size of the earth became visible in it, even under a powerful microscope. Their general shape is similar to that which Sir William Herschel assigned to the galactic system, and it was this that originally led him to regard them as "island universes" similar to the galactic system. We shall see later how far his conjecture has been confirmed by recent research.

The Distances of the Stars

The year 1838 provides our next landmark; it is the year in which the distance of a star was first measured.

In the second century after Christ, Ptolemy had argued that if the earth moved in space, its position relative to the surrounding stars must continually change. As the earth swung round the sun, its inhabitants would be in the position of a child in a swing. And, just as the swinging child sees the nearer trees, persons and houses oscillating rhythmically against a remote background of distant hills and clouds, so the inhabitants of the earth ought to see the nearer stars continually changing their position against their background of more distant stars. Yet night after night the constellations remained the same, or so Ptolemy argued; the same stars circled eternally in the same relative positions around the pole, and conspicuous groups of stars such as the seven stars of the Great Bear, the Pleiades or the constella-

PLATE IV

Yerkes Observatory.

The Great Nebula (*M* 31) in Andromeda.

PLATE V

Mt. Wilson Observatory.

The Nebula N.G.C. 891 in Andromeda seen edge-on.

tion of Orion shewed no signs of change. For aught the unaided human eye could tell, the stars might be spots of luminous paint on a canvas background, with the earth as the unmoving pivot around which the whole structure swung.

In opposition to this, the Copernican theory of course required that the nearer stars should be seen to move against the background of the more distant stars, as the earth performed its yearly journey round the sun. Yet year after year, and even century after century, passed without any such motion being detected. The old Ptolemaic contention that the earth formed the fixed centre of the universe might almost have regained its former position, had it not been that various lines of evidence had begun to shew that even the nearest stars were necessarily very distant, so distant, indeed, that their apparent want of motion need cause no surprise. The child in a swing cannot expect to have optical evidence of its motion if the nearest object it can see is twenty miles away.

Very few stars appear brighter than Saturn at its brightest; it looks about as bright as Altair, the eleventh brightest star in the sky. Yet Saturn shines only by the light it reflects from the sun, and its distance from the sun is such that it receives only about one part in 10,000 million of the light emitted by the sun. If, as Kepler and others had maintained, Altair was essentially similar to the sun, it would probably be of about the same candle-power as the sun, and so would give out about 10,000 million times as much light as Saturn. In other words, if Altair were placed in the position of Saturn, it would appear about 10,000 million times as bright as Saturn. The fact that Altair and Saturn appear about equally bright in the sky can only mean that Altair is 100,000 times as distant as Saturn.* This argument is essentially identical with one which Newton gave in his *System of the World* to shew that even the brightest stars, such as Altair, must be very distant indeed.

And such has proved to be the case. All efforts to discover the apparent swinging motion of the stars—"parallactic motion," as it is technically called—which results from the earth's orbital motion failed until 1838, when three astronomers, Bessel, Henderson, and Struve, almost simultaneously detected the parallactic motions of the three stars, 61 Cygni, α Centauri and α Lyrae respectively. The amount of their parallactic motion made it possible to calculate the distances of the stars, so that the inhabitants of the earth were not only placed in possession of definite ocular proof that they were swinging round the sun, but from the visible effects of this swing they were able to compute the distances of the nearer stars. The calculated values were not accurate when judged by modern standards, but they provided the first definite estimates of the scale on which the universe is built.

Let us pause for a minute to consider how this scale is built up. The first step is to select a convenient base-line a few miles in length on the surface of the earth, and to measure this in terms of standard yards or metres. Starting out from this base-line, a geodetic survey maps out a long narrow strip of the earth's surface, preferably running due north and south. The difference of latitude at the two ends is then measured by astronomical methods, as for instance by noticing the difference in the altitude of the pole-star at the two places. As the length of the strip is already known in miles, this immediately gives the dimensions of

* For the apparent brightness of an object falls off as the inverse square of its distance, and $(100,000)^2 = 10,000,000,000$.

EXPLORING THE SKY

the earth. According to Hayford (1909), the earth's equatorial radius is 6378.388 kilometres, or 3963.34 miles, its polar radius being 6356.909 kilometres or 3949.99 miles.

The next step is to determine the size of the solar system in terms of that of the earth. When the sun is eclipsed by the moon, the time at which the moon first begins to cover the sun's disc is different for different stations on the earth's surface, and the observed differences of time enable us to measure the moon's distance in terms of known distances on the surface of the earth. In this way the mean distance of the moon is found to be 384,403 kilometres or 238,857 miles. In the same way the transit of the planet Venus across the disc of the sun provides an opportunity for determining the scale of the solar system in terms of the dimensions of the earth. The asteroid Eros provides still better opportunities. The Paris Conference (1911) adopted 149,450,000 kilometres, or 92,870,000 miles, as the most likely value for the mean distance of the earth from the sun. The next and final step, which the year 1838 saw accomplished, is that of using the diameter of the earth's orbit as base-line, and determining the distances of the stars.

The first step, from the standard yard or metre to the measured base-line on the earth's surface, involves an increase of several thousand-fold in length. The increase involved in the next step, from the base-line to the earth's diameter, is again one of thousands. And again the next step, from the diameter of the earth to that of the earth's orbit, involves an increase of thousands. But the last step of all, from the earth's orbit to stellar distances, involves a million-fold increase.

Recent measurements shew that the nearest stars are at almost exactly a million times the distances of the nearest planets. At its nearest approach to the earth, Venus is 26 million miles distant, while the nearest star, Proxima Centauri is 25,000,000 million miles away; this latter star is a faint companion of the well-known bright star α Centauri in the southern hemisphere. The distances of the planets when at their nearest, and of the nearest stars, are shewn in the following table.

As it is almost impossible to visualise a million, the mere statement that the stars are a million times as remote as the planets gives only a feeble indication of the immensity of the gap that divides the solar system from its nearest neighbours in space. Perhaps the apparent fixity of the stars gives a more vivid impression.

Planets		Stars		
Name	Distance (miles)	Name	Distance (miles)	Distance (light-years)
Venus	26,000,000	Proxima Centauri	25,000,000 million	4.27
		α Centauri		4.31
Mars	35,000,000	Munich 15040	36,000,000 "	6.06
		Wolf 359	47,000,000 "	8.07
Mercury	47,000,000	Lalande 21185	49,000,000 "	8.33
		Sirius	51,000,000 "	8.65

The earth performs its yearly journey round the sun at a speed of about 18½ miles a second, which is about 1200 times the speed of an express train. The sun moves through the stars at nearly the same rate—to be precise, at about 800 times the speed of an express train. And, broadly speaking, the nearer planets and the majority of the stars move with similar speeds. We shall not obtain a bad approximation to the truth if we imagine that all astronomical bodies move with exactly equal speeds, let us say, to fix our thoughts, a speed equal to 1000 times the speed of an express train. The distances of astronomical objects are now betrayed by the speed with which they appear to move across the sky—the slower their apparent motion the greater their distances, and *vice versa.* Now the planets move across the sky so rapidly that it is quite easy to detect their motion from night to night and even from hour to hour; the stars move so slowly that, except with telescopic aid, no motion can be detected from generation to generation, or even from age to age. Even the conspicuous constellations in the sky, which on the whole are formed of the nearer stars, have retained their present appearance throughout the whole of historic times. The contrast between the planets which change their positions every hour, and the stars which fail to shew any appreciable change in a century, gives a vivid impression of the extent to which the stars are more distant than the planets.

It is far more difficult to visualise the actual distances of the stars. The statement that even the nearest of them is 25,000,000 million miles away hardly conveys a definite picture to the mind, but we may fare better with the alternative statement that the distance is 4.27 light-years—that is to say the distance that light, travelling at 186,000 miles a second, takes in 4.27 years to traverse.

Light travels with the same speed as wireless signals because both are waves of electric disturbance. Incidentally this speed is just about a million times that of sound. The enormous disparity in the speeds of sound and of electric waves is vividly brought out in the ordinary process of broadcasting. When a speaker broadcasts from London his voice takes longer to travel 3 feet from his mouth to the microphone as a sound wave, than it does to travel a further 560 miles to Berlin or Milan as an electric wave. Wireless listeners in Australia hear the music of a broadcast concert sooner than an ordinary listener at the back of the concert hall who relies on sound alone; they hear it a fifteenth of a second after it is played. Yet light, or wireless waves travelling with the same speed as light, takes 4.27 years to reach the nearest star, so that the inhabitants of Proxima Centauri would be over four and a quarter years late in hearing a terrestrial concert. And in time we shall have to consider other and even more distant stars which terrestrial music would not yet have reached had it started on its journey before the Norman Conquest, before the Pyramids were built, even before man appeared on earth.

The Photographic Epoch

If we were only allowed to select one more landmark in the progress of astronomy, we might well choose the application of photography to astronomy in the closing years of the nineteenth century; this opened the floodgates of progress more thoroughly than anything else had done since the invention of the telescope. Hitherto the telescope, after collecting and bending rays of light from the sky, had projected the concentrated beam of light through the pupil of the human eye on to the retina; in future it was to project it on to the incomparably more sensitive photographic plate. The eye can retain an impression only for a fraction of a second; the photographic plate adds up all the impressions it receives for hours or even days, and records them practically for ever. The eye can only measure distances between astronomical objects by the help of an intricate machinery of cross-wires, screws and verniers; the photographic plate records distances automatically. The eye, betrayed by preconceived ideas, impatience or hope, can and does

PLATE VI

Mt. Wilson Observatory.

The "Horse's Head" in the Great Nebula in Orion.

make every conceivable type of error; the camera cannot lie.

And so it comes about that if we try to pick out landmarks in twentieth-century astronomy we find that, in a sense, it consists of nothing but landmarks; the slow, arduous methods of conquest of the nineteenth century have given place to a sort of gold-rush in which claims are staked out, the surface scratched, the more conspicuous nuggets collected, and the excavation abandoned for something more promising, all with such rapidity that any attempt to describe the position is out of date almost before it can be printed. We can only attempt a general impression of the new territory, and with this will be inextricably mixed a discussion of old territory seen in the light of new knowledge.

Groups of Stars and Binary Systems

A glance at the sky, or, better, at a photograph of a fragment of the sky, suggests that, in the main, the stars are scattered at random over the sky, except for the concentration of faint stars in and towards the Milky Way, which we have already considered. Any small bit of the sky does not look very different from what it would if bright and faint stars had been sprinkled haphazard out of a celestial pepperpot.

Yet this is not quite the whole story. Here and there groups of conspicuous stars are to be seen, which can hardly have come together purely by accident. Orion's belt, the Pleiades, Berenice's hair, even the Great Bear itself, do not look like accidents, and in point of fact are not. It is the existence of these natural groups of stars that lies at the root of, and justifies, the division of the stars into constellations. We shall explain later how the physical properties of the stars are studied; for the present it is enough to remark that physical study confirms the suspicion that groups such as those just mentioned are, generally speaking, true families, and not mere accidental concourses, of stars. The stars of any one group, such as the Pleiades, not only shew the same physical properties, but also have identical motions through space, thus journeying perpetually through the sky in one another's society. As such a group of stars are both physically similar, and travel in company, they might appropriately be described as a family of stars. The astronomer, however, prefers to call them a "moving cluster."

These families are of almost all sizes, the smallest and commonest type consisting of only two members. After this the next commonest type consists of three members; our nearest three neighbours in space, Proxima Centauri and the two stars of α Centauri, form such a triple system. Then come systems of four, five and six members, and so on indefinitely.

Let us first turn our attention to families consisting of only two members—"binary systems," as they are generally called. Even if the stars had been sprinkled on to the sky at random out of a pepperpot, the laws of chance would require that in a certain number of cases, pairs of stars should appear very close together. And a study of a photograph of any star-field shews that a large number of such close pairs actually exist. The number is, however, greater than can be explained by the laws of chance alone. Some pairs of stars may be close together by accident, but a physical cause is needed to account for the remainder. We can unravel the mystery by photographing the field at intervals of a few years and comparing the various results obtained. Some of the stars which originally appeared as close pairs will be found to move steadily apart. These are the pairs of stars which, although they appeared close together in the sky, were not so in space; one star merely happened to be almost exactly in line with the other as seen from the earth. Other pairs do not break up with the passage of time; the two components change their relative positions but never become completely separated. In the simplest case of all one star may be found to describe an approximately circular orbit about the other, just as the earth does round the sun, and the moon round the earth, and for precisely the same reason: gravitation keeps them together.

The Law of Gravitation. Drop a cricket ball from your hand and it falls to the ground. We say that the cause of its fall is the gravitational pull of the earth. In the same way, a cricket ball thrown into the air does not move on for ever in the direction in which it is thrown; if it did it would leave the earth for good, and voyage off into space. It is saved from this fate by the earth's gravitational pull which drags it gradually down, so that it falls back to earth. The faster we throw it, the further it travels before this occurs; a similar ball projected from a gun would travel for many miles before being pulled back to earth.

The law governing all these phenomena is quite simple. It is that the earth's gravitational pull causes all bodies to fall 16 feet earthward in a second. This is true of all bodies which are free to fall, no matter how they are moving; every body which is not in some way held up against gravitation is 16 feet lower at the end of any second than it would have been if gravitation had not acted through that second.

To illustrate what this means, let the big circular curve $B'A'C'$ in fig. 2 represent the earth's surface, and imagine that a shot is fired horizontally from *A*, the top of an elevation *AA'*. If the shot were not pulled earthwards by gravitation, it would travel indefinitely along the line *AB* out into space. If *AB* is the distance it would travel in a second under these imaginary conditions, the end of a second's actual flight does not find it at *B*, but at a point 16 feet nearer the earth, gravitation having pulled it down this 16 feet during its flight. For instance, if *BB'* in fig. 2 should happen to be 16 feet, the shot would strike the earth at *B'* after a flight of precisely one second.

As another example, let us suppose that the 16-foot fall below *B* does not drag the shot down to earth but only to

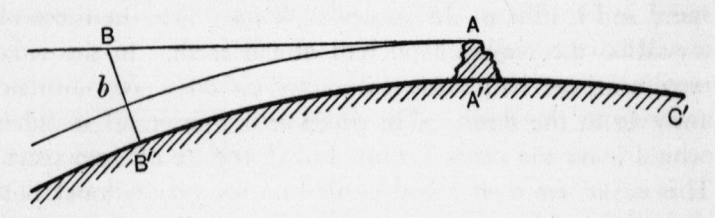

FIG. 2.

a point *b*, which is at precisely the same height above the earth's surface as the point *A* at which the shot started. If gravitation were not acting, so that the shot travelled along the line *AB*, its height above the earth would continually increase. Actually in the case we are now considering, gravitation pulls the shot down at just such a rate as to neutralise the increase of height which would otherwise occur, so that the height of the shot neither increases nor decreases; it neither flies off into space nor drops to earth, but continues to describe circles round the earth indefinitely.

A simple geometrical calculation shews that for the distance *Bb* to be 16 feet, the distance *AB* travelled in one second must be 25,880 feet or 4.90 miles.* Thus if we could fire a shot horizontally with a speed of 4.90 miles a second, it would describe endless circles round the earth, the earth's gravitational pull exactly neutralising the natural tendency of the shot to fly away along the straight line AB.

In 1665 Newton began to suspect that this same gravitational pull might be the cause of the moon describing a circular orbit around the earth instead of running away at a tangent into space. The moon's distance from the earth's centre is 238,857 miles, or 60.27 times the radius of the earth. As the moon describes a circle of this size every month (27 days, 4 hours, 43 minutes, 11.5 seconds), we can calculate that its speed in its orbit is 2287 miles an hour. After one second it will have travelled 3350 feet, and if it kept to a strictly rectilinear course this would carry it 0.0044 feet further away from the earth. Thus to keep in an exact circular orbit around the earth, it must fall 0.0044 feet in a second. This is far less than a body falls in a second at the earth's surface, but Newton conjectured that the force of gravity

* Let C be the centre of the earth, and bCD the diameter through b.
Then $BA^2 = Bb \times BD$, where $Bb = 16$ feet, and BD, which is 16 feet more than the earth's diameter $= 41,900,000$ feet. From this we readily calculate

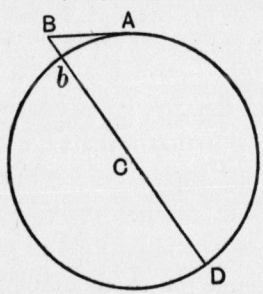

FIG. 3.

that $BA = 25,880$ feet. This calculation of course neglects the height of the hill AA' by comparison with the earth's diameter.

must weaken as we recede from the earth's surface. Actually a body at the earth's surface falls 3632 times as fast as the moon's earthward fall in its orbit. Now 3632 is the square of 60.27 (or $3632 = 60.27 \times 60.27$), whence Newton saw that the moon's fall would be of exactly the right amount if the force of gravity fell off as the inverse square of the distance—that is to say, if it decreased just as rapidly as the square of the distance increased. As we shall see later, astronomical observation confirms the truth of this law in innumerable ways. This led Newton to put forward his famous law of gravitation according to which the gravitational pull of any body, such as the earth, falls off inversely as the square of the distance from the body.

Professor C. V. Boys and others have measured the gravitational pull which a few tons of lead exert in the laboratory, and, with this knowledge, it is easy to calculate how many tons the earth must contain so as to exert its observed gravitational pull on bodies outside it. It is found that the earth's weight must be just under six thousand million million million tons,* or, as we shall write it, 6×10^{21} tons.†

Just as the earth's gravitational pull keeps the moon perpetually describing circles around it, so the sun's gravitational pull keeps the earth and all the other planets describing circles around the sun. Knowing the distance of any planet from the sun, and also its speed in its orbit, we can

---

* Here, as throughout the book, we use the French or metric ton of a million grammes or 2204.5 lbs. The English ton of 2240 lbs. is equal to 1.0160 French tons.

† The notation 6×10^{21} stands for the number formed by a 6 followed by 21 zeros, this shorthand notation being essential, in the interests of brevity, in discussing astronomical numbers. A million is 10^6, a million million is 10^{12} and so on.

A similar notation is needed to express very small numbers. The expression 10^{-21} is written for $1/10^{21}$ and so on. Thus 6×10^{-6} stands for 6/1,000,000 or 0.000006.

PLATE VII

Observatory.

The Trifid Nebula (*M* 20) in Sagittarius.

calculate the distance this planet falls towards the sun in a second. This tells us the amount of the sun's gravitational pull, and from this we can calculate that the sun's weight must be about 332,000 times the weight of the earth, or almost exactly 2×10^{27} tons. Whichever of the planets we use, we obtain exactly the same weight for the sun. This not only gives us confidence in our result, but incidentally it also provides striking confirmation of the truth of Newton's law of gravitation, for if this law were inexact or untrue, the different planets would not all tell exactly the same story as to the sun's weight. Einstein has recently shewn that the law is not absolutely exact, but the amount of inexactness is inappreciable except for the nearest planet, Mercury, and even here it is so exceedingly small that we need not trouble about it for our present purpose.

Just as we can weigh the sun and earth by studying the motion of a body gripped by their gravitational pull—or "in their gravitational fields," as the mathematician would say—so we can weigh any other body which keeps a second small body moving round it by its gravitational attraction. The motions of Jupiter's satellites make it possible to weigh Jupiter; its weight is found to be about 1.92×10^{24} tons, which is 317 times that of the earth, although only 1/1047 of that of the sun. Similarly the weight of Saturn is found to be 5.71×10^{23} tons or about 94.9 times that of the earth.

Weighing the Stars. And now we come to a striking application of the principles just explained—when we observe two stars in the sky describing orbits about one another, we can weigh the stars from a study of their orbits. Generally the problem is not quite so simple as those we have just discussed. For its adequate treatment, we must once again levy toll on the mathematical work of Newton.

We have seen that a projectile fired horizontally with a speed of 4.90 miles a second, would describe endless circles round the earth. What would happen if it were fired in some other direction and with some other speed?

The answer was provided by Newton. He shewed that when a small body moves in any way whatsoever under the gravitational force of a big body, its orbit is always an ellipse —a sort of pulled out circle or oval curve * (fig. 4, p. 45). Previous to this Kepler had found that the actual paths of the planets round the sun were not exact circles but ellipses, although for the most part ellipses which did not differ greatly from circles; they are what the mathematician calls "ellipses of small eccentricity." This provides still further confirmation of Newton's law of gravitation, for it can be proved that if the force of gravitation falls off in any way other than according to Newton's law of the inverse square of the distance, the orbits of the planets will not be elliptical.

When the astronomer studies the motions of a binary star in the sky, he generally finds that the two components do

---

* The simplest definition of an ellipse is that it is the curve drawn by a moving point P which moves in such a way that the sum of its distances PS, PT from two fixed points S, T remains always the same. In practice we can most easily draw an ellipse by slipping an endless string $SPTS$ round two drawing pins S, T stuck into a drawing board. Stretch the string tight with a pencil at P, and on letting the pencil move round, keeping the string always tight, we shall draw an ellipse. If the pins S, T in the drawing board are placed near to one another the curve described by the pencil P is nearly circular. The ratio of the distance ST to the length of the remainder of the string $SP+PT$ is called the "eccentricity" of the ellipse; it is necessarily less than unity, because two sides of a triangle are together greater than the third side.

In the limiting case in which the eccentricity is made zero, the ellipse becomes a circle. If the eccentricity is nearly as large as unity, the ellipse is very elongated. All the different shapes of ellipses are obtained by letting the eccentricity change from 0 to 1, and these represent all the different shapes of orbit that a small body can describe around a heavy gravitating mass. The points S, T are called the foci of the ellipse, and the big attracting body always occupies one or other of the two foci of the ellipse.

not move in circles about one another but in ellipses.* Once again, Newton's law is confirmed, and we are entitled to assume that the forces which keep binary stars together are the same gravitational forces as keep the moon from running away from the earth, or the planets from the sun. By a study of these ellipses it becomes possible to weigh the stars. If one of the component masses were enormously

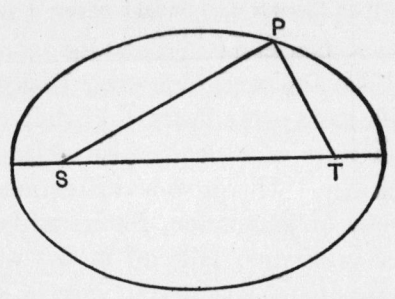

Fig. 4.—The oval curve is an ellipse; the points S, T are its "foci."

heavier than the other, the former would stand still while the lighter component described an ellipse around it, the motion being essentially similar to that of a planet around the sun. Such cases are not observed in actual binary stars because the two components are generally comparable in weight, and this brings new complications into the question. There is no need to enter into mathematical details here. Suffice it to say that neither star stands still; the two components describe ellipses of different sizes, and from a study of these two ellipses the weights of both the components can be determined.

* What he actually observes is the "projection" of the orbit on the sky, but it is a well-known theorem of geometry that the projection of an ellipse is always an ellipse.

The following table shews the result of weighing the four binary systems nearest the sun in this way, the sun's weight being taken as unity:

Stellar Weights

Binary systems near the sun

Star	Distance in light years from the sun	Weights of components in terms of sun's weight	Luminosity see (p.47)
α Centauri A	4.31	1.14	1.12
" B		0.97	0.32
Sirius A	8.65	2.45	26.3
" B		0.85	0.0026
Procyon A	10.5	1.24	5.5
" B		0.39	0.00003
Kruger 60 A	12.7	0.25	0.0026
" B		0.20	0.0007

We see that the weights of these stars do not differ greatly from that of the sun, although naturally the whole of space provides a greater range than the four stars of our table which happen to be near the sun. But even in the whole of space, no star whose weight is known with any accuracy has a weight less than Kruger 60 B, although at the other end of the scale there are many stars with far greater weights than any in our table. Of stars whose weights are known with fair accuracy, the star H.D. 1337 (Pearce's star) is the weightiest, its two components being respectively 36.3 and 33.8 times as heavy as the sun. Plaskett's star B.D. 6° 1309. is certainly heavier still, its components weighing at least 75 and 63 times as much as the sun, and probably more; the exact weights are not known (see p. 52 below). The system 27 Canis Majoris consists of four stars, whose combined weight, according to the evidence at present available, appears to be at least 940 times that of the sun, but we may properly exercise a certain amount of caution before accepting a figure so far outside the usual run of stellar weights.

The average constituent star in the above very short table has 0.94 times the weight of the sun, so that our sun appears to be of rather more than average weight, and this is confirmed by a more extensive study of stellar weights.

We might have expected *à priori* that the stars would prove to have all sorts of weights, for there is no obvious reason why stars should not exist with weights millions of times that of the sun, or again with weights only equal to that of the earth or less. Actually we find that the weights of the stars are mostly fairly equal, very few stars having weights greatly dissimilar from that of the sun. This seems to indicate that a star is a definite species of astronomical product, not a mere random chunk of luminous matter.

Luminosity. The last column of the table on p. 46 gives the "luminosities" of the stars, which means their candlepower as lights, that of the sun being taken as unity. For instance the entry of 26.3 for Sirius means that Sirius, regarded as a lighthouse in space, has 26.3 times the candlepower of the sun. The luminosities of the stars shew an enormously greater range than their weights. In a general way the heaviest stars are found to be the most luminous, as we should naturally expect, but their luminosity is out of all proportion to their weight. The heavier component of Sirius has only 2.9 times the weight of the lighter component, but 10,000 times its luminosity. Again, in the system of Procyon the heavier component has 3.2 times the weight, but 180,000 times the luminosity, of the lighter component. It appears to be an almost universal law that the candlepower per ton is far greater in heavy stars than in light.

This is one of the central and, at first sight, one of the most perplexing facts of physical astronomy: it is so fundamental and so pervading that no view of stellar mechanism can be accepted which fails to explain it.

Spectroscopic Velocities. When a star's distance is known, its motion across the sky tells us its speed in a direction at right angles to the line along which we look at it—i.e. across the line of sight—but provides no means of discovering its speed along this line. We cannot see the motion of a body which is coming straight towards us, and a star moving at a million miles a second in a direction exactly along the line of sight, would yet appear to be standing still in the sky. To evaluate velocities along the line of sight, the astronomer calls in the aid of the spectroscope.

All light is a blend of lights of different colours, and just as Newton, with his famous prism, analysed sunlight into all the colours of the rainbow, so the spectroscope analyses the light from a star, or indeed from any source whatever, into its various constituent colours. The instrument spreads out the analysed light into a strip of light of continuously graduated colour, which is described as a "spectrum." In this the colours are the same, and are found to be arranged in the same order, as in the rainbow, running from violet through green and orange to red. There is a physical reason for this. We shall see later (p. 115) that light consists of trains of waves—like the ripples which the wind blows up on a pond—and that the different colours of light result from waves of different lengths, red light being produced by the longest waves, and violet light by the shortest. The colours in the spectrum occur in the order of their wave-lengths, from the longest (red) to the shortest (violet). In the typical stellar spectrum certain short ranges of colour or

PLATE VIII

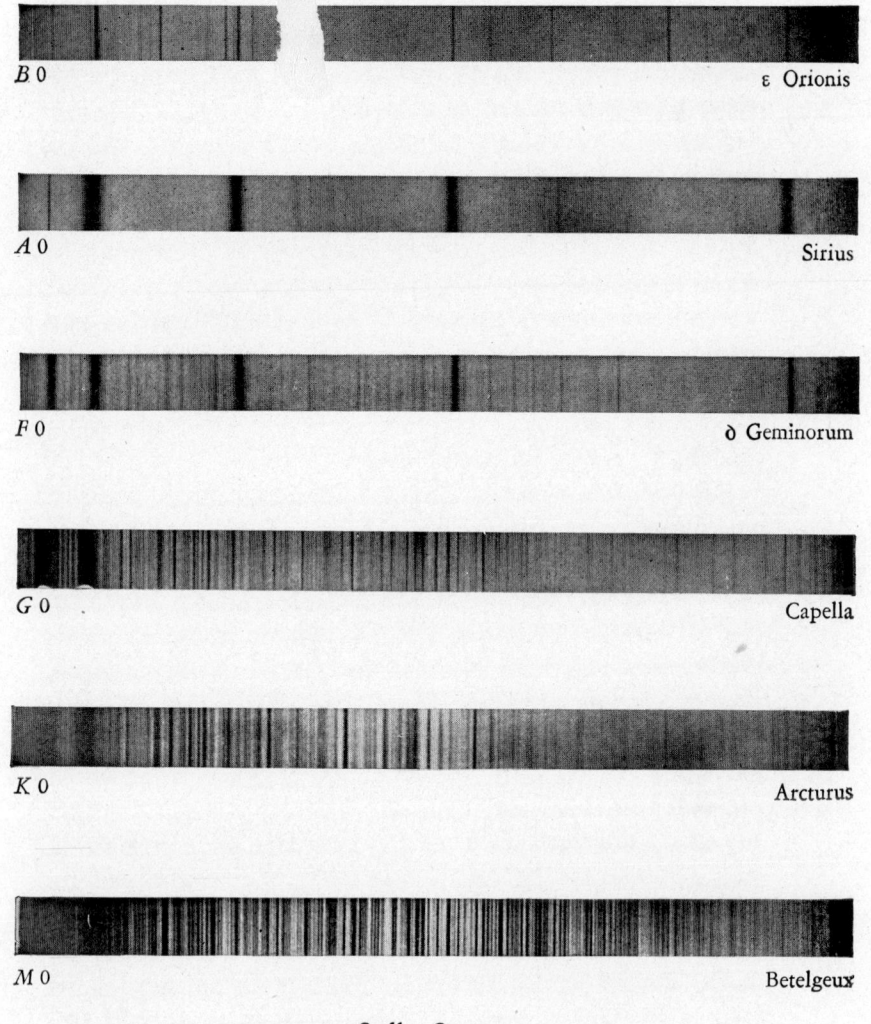

B 0   ε Orionis

A 0   Sirius

F 0   δ Geminorum

G 0   Capella

K 0   Arcturus

M 0   Betelgeux

Stellar Spectra.
(The spectral types are indicated on the left.)

wave-length are generally missing, for reasons we shall discuss later (p. 118), so that the spectrum appears to be crossed by a number of dark lines or bands, thus forming a pattern rather than a continuous gradation of colours. Examples of stellar spectra are shewn in Plate VIII. It is frequently convenient to classify stars by the type of spectra they emit. It is found that spectra can, in the main, be arranged in a single continuous sequence, and their usual classification is by a sequence of letters, *B, A, F, G, K, M* with decimal subdivisions. Spectral types are indicated on the left in Plate VIII.

When the light received from a star is analysed in a spectroscope, the pattern of lines or bands may be found to be shifted bodily in one direction or the other. If the shift is towards the red end of the spectrum, the light emitted by the star is reaching us in a redder state than that in which it ought normally to be, and as red light has the longest wave-length, this means that every wave of light is longer—more drawn out—than normal. We conclude that the star is receding from us. In the same way, if the spectral pattern is shifted toward the violet end of the spectrum, we know that the star must be approaching us.* From the observed amount of the displacement of the spectrum we can calculate the star's actual speed along the line of sight, and the calculation is surprisingly simple. If each line or band in a spectrum is found to represent a wave-length a hundredth of one per cent. longer than that usually associated with it, then the star's speed of recession is a hundredth of one per cent. of the velocity of light, or 18.6 miles a second—and similarly for all other displacements.

* The shift of a spectrum resulting from the motion of the body which emits it is generally described as the Doppler-effect.

Spectroscopic Binaries. As the two components of a binary system are generally moving with different speeds, the normal spectrum of a binary system consists of two distinct superposed spectra, the two spectra shewing different shifts which correspond to the speeds of the two components. From the observed orbits of the two components of a binary system, an astronomer might proceed to calculate with what speeds these components would move in the direction of the line of sight, and could then predict to what extent the two spectra ought to be displaced if the light from the system were analysed in a spectroscope; the spectroscope would of course confirm his prediction.

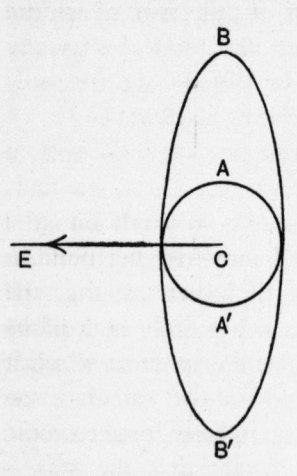

FIG. 5.—The little orbit *AA'* and the big orbit *BB'* give the same velocities along the line of sight *CE*.

It is more instructive to imagine the reverse process. Suppose that on analysing the light from a star, the astronomer obtains a composite spectrum in which two distinct spectra shift rhythmically backwards and forwards about their normal position. The fact that there are two spectra tells him that he is dealing with a binary system; if the rhythmic shift repeats itself every two years, he knows that its orbit takes two years to complete. He studies the star by direct vision and finds it is a binary system in which the constituents revolve about one another every two years.

He examines another spectrum, and finds that it shifts rhythmically every two days. On looking directly at this star he can only see a single point of light. There must, of course, be two stars, but the mere fact that they get around one another in so short a time as two days proves that they must be very close to one another, and he need feel no surprise that his telescope has failed to separate the image into two distinct points of light. Systems of this kind, which the spectroscope shews to be binary, but the telescope usually shews as a single point of light, are called "spectroscopic binaries." Over a thousand such systems are known.

If the astronomer tries to construct the orbit of such a system from the spectroscopic observations alone, he finds himself in difficulties. His observations only tell him the velocities along the line of sight, and these depend both on the actual speed and on the degree of foreshortening; the same velocity may arise either from a big orbit in a plane nearly at right angles to the line of sight, or from a much foreshortened little orbit. It is impossible to calculate the actual orbit or the weights of the stars from spectroscopic observation alone.

Eclipsing Binaries. There is one exception. Suppose that a star's light is seen to diminish in amount at regular intervals and subsequently to return to its original strength. The obvious interpretation of the diminution of light is that one component of the system is eclipsing the other, and this can only happen if the orbit is so completely foreshortened that its plane passes through, or at least very close to, the earth. In such a case it is possible to reconstruct the whole orbit, and thence to calculate the weights of the two components. Not only so, but the length of time during which the eclipses last tells us the actual sizes of the two components, so that it is possible to draw a complete picture of the system.

Diagrams of the dimensions and orbits of two typical eclipsing binaries are shewn in fig. 6; these are drawn to the same scale, this being indicated by the small circle representing the sun.

When no eclipse occurs in a spectroscopic binary, we do not know how much foreshortening to allow for, but we can obtain a general idea of the weights of the components by assuming an average degree of foreshortening. If we assume

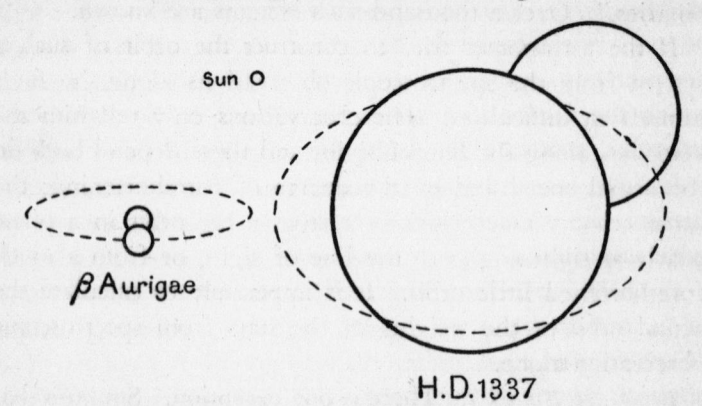

Sun O

β Aurigae

H.D. 1337

FIG. 6.—Components and orbits of Eclipsing Binaries.

different degrees of foreshortening in turn, we shall find that the computed weights come out least when the plane of the orbit is assumed to pass through the earth—i.e. when the orbits are computed as though the system were an eclipsing one. Thus although we cannot discover the actual weights of the components of a non-eclipsing binary, we can always state limits above which they must lie, namely the weights computed as though the system were an eclipsing one. In this way, we know that the two components of Plaskett's star must have more than 75 and 63 times the weight of the sun.

Variable Stars

The majority of stars shine with a perfectly steady light, so that we can say that a star is of so many candle-power. The sun, for instance, emits a light of 3.23×10^{27} candle-power.

Yet there are classes of exceptional stars in which the light flickers up and down. In some, as in the eclipsing binaries just described, the light-fluctuations are quite regular, repeating themselves with such precision that the stars might well be used as timekeepers. In others the fluctuations, though not perfectly regular, are nearly so, while still others exist in which the fluctuations appear at present to be completely irregular, although no doubt the changes in these will be reduced to law and order in due course. For our present discussion, the various types of irregular variables are not of great importance.

Cepheid Variables. The really interesting stars are those of a certain class of regular variable, generally called "Cepheid variables," after their prototype δ Cephei. The physical nature of these stars and the mechanism of their light-fluctuation is still far from being understood; competing theories are in the field which we need not discuss at this stage (see p. 212 below).

Whatever their mechanism may be, observation shews that these stars possess a certain definite property, which proves to be of the utmost value. This being so, we may accept it gratefully without troubling as to its why and wherefore. The perfectly regular light fluctuations of the eclipsing binaries would make them suitable for time-keepers even though we did not understand the mechanism behind these fluctuations. In the same way the fluctuations of Cepheid variables have a quality which makes them valuable as measuring-rods with which to survey the distant parts of the universe. In brief, this property is that we can deduce the intrinsic brightness of these stars, and so their distances, from their observed light-fluctuations.

The light-fluctuations are so distinctive as to make the stars easy of detection. There is a rapid increase of light, followed by a slow gradual decline; then again the same rapid increase and slow decline as before. It is as though someone were throwing armfuls of fuel onto a bonfire at perfectly regular intervals.

One other class of variable stars, generally known as "long-period variables" shews somewhat similar light fluctuations, but the two classes are easily distinguished by their very different periods of light-fluctuation. The Cepheid variable completes its cycle in a time which may be a few hours, or may be days or weeks, but is never more than about a month, whereas the long-period variable generally requires about a year.

Fig. 7 shews the light-curves of typical variable stars of the different classes. In each diagram the progress of time is represented by motion across the page from left to right; the higher the fluctuating curve is above the horizontal line at any instant, the brighter the star at that instant.

Out near the boundary of the galactic system is a cluster of stars known as the Lesser Magellanic Cloud (Plate XXI, p. 204), in which Cepheid variables occur in great profusion. In 1912, Miss Leavitt of Harvard found that the light of the brighter Cepheids in this cloud fluctuated more slowly than the light of the fainter ones. Whatever was responsible for turning the stellar lights up and down, acted more rapidly for feeble than for brilliant lights. The apparent brightnesses of a number of Cepheids at varying distances would of

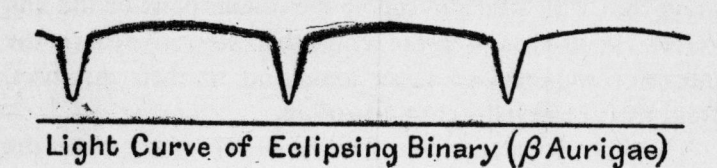

Light Curve of Eclipsing Binary (β Aurigae)

Light Curve of Irregular Variable (RS Ophiuchi)

Light Curve of Cepheid Variable (ν Lacertae)

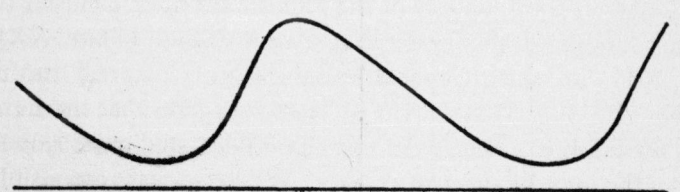

Light Curve of Long Period Variable (o Ceti)

FIG. 7. Light Curves of typical Variable Stars of different classes.

course depend only in part on their intrinsic brightness or candle-power, but the stars in the Magellanic Cloud are all, nearly enough, at the same distance from the earth. Thus differences in the apparent brightnesses of stars in this cloud must represent real differences of intrinsic brightness, and Miss Leavitt's discovery could be stated in the form that the period of light-fluctuation of a Cepheid depended on its candle-power. Although this was only proved for the Cepheids in the Magellanic Cloud, it must be true for all Cepheids wherever they are, for it is inconceivable that we could make a star's light fluctuate more slowly or more rapidly merely by altering its distance from us—by ourselves receding from it, in fact.

Professor Hertzsprung of Leiden and Dr. Shapley, then of Mount Wilson Observatory, were quick to seize upon the implications of this discovery. If two Cepheids A, B in different parts of the sky are found to fluctuate with equal rapidities, then their intrinsic candle-powers must be equal. Thus any difference in their apparent brightness must be traceable to a difference in their distances from us. If A looks a hundred times as bright as B, then B must be at ten times the distance of A. In the same way, a third Cepheid C may prove to be at ten times the distance of B. We now know that C is a hundred times as remote as A. And if D can be found ten times as distant as C, we know that D is a thousand times as remote as A. So we can go on constructing and ever extending our measuring-rod; there is no limit until we reach distances so great that even Cepheid variables, which are exceptionally bright stars, fade into invisibility.

So far we have only considered the comparative distances of Cepheids. The absolute distances of many of the nearer Cepheids have, however, been determined by the parallactic method already explained—i.e. by measuring their apparent motion in the sky, resulting from the earth's motion round the sun. Taking any one of these stars as our original Cepheid A, we can step continually from one Cepheid to another, and so calculate the absolute distances of all the Cepheid variables in the sky.

In this way the observed relation between the period of fluctuation and the brightness of Cepheid variables—commonly known as the "period-luminosity law"—can be made to provide a scale on which the absolute luminosity, or candle-power, of a Cepheid can be read off directly from the observed period of its light-fluctuations. The Cepheid variables may be regarded as lighthouses set up in distant parts of the universe. We can recognise them, just as a sailor recognises lighthouses, by the quality of their light. We can read off their candle-power from the period of their observed light-fluctuations as easily as the sailor could read off the candle-power of a lighthouse from an Admiralty chart. The apparent brightness of the Cepheid informs us as to its distance from us.*

It would be difficult to over-estimate the importance of

* For instance, Cepheids whose light fluctuates in a period of 40 hours have approximately a luminosity 250 times that of the sun, and so are of 8×10^{29} candle-power; a period of ten days indicates a luminosity 1600 times that of the sun, or a candle-power of 5.17×10^{30}, and so on. If a star in a distant astronomical object is observed to fluctuate with a period of 10 days, and the quality of its fluctuations shew it to be a Cepheid variable, we know that its actual candle-power must be 5.17×10^{30}. Its apparent brightness is observed to be that of a star of, say, magnitude 16, which, stripped of technicalities, means that we receive as much light from it as from a single candle at a distance of 10,000 miles. The difference between one candle and 5.17×10^{30} candles accordingly corresponds to the difference between 10,000 miles and the distance of the object in question, whence, since light falls off as the inverse square of the distance, we calculate that the distance of the object must be

$$\sqrt{5.17 \times 10^{30}} \times 10,000 \text{ miles}$$

or about 3,600,000 light-years.

all this to modern astronomical science. It means that a method has been found for surveying, if not the whole of the universe, at least those parts of it in which Cepheid variables are visible. Actually this last reservation is unimportant, for Cepheid variables are very freely scattered in space. Naturally the method is of most value for the exploration of the most distant parts of the universe; here it achieves triumphant success where other methods fail completely. The parallactic method begins to fail when we try to sound distances of more than about a hundred light-years. The apparent path in the sky, which a star at this distance describes, in consequence of the earth's motion round the sun, is of the size of a pin-head two miles away. With all their refinements, modern instruments find it difficult enough to detect so small a motion as this, and it is practically impossible to measure it with accuracy. The "period-luminosity" law measures the distances of objects up to a million light-years away, with a smaller percentage of error than is to be expected in the parallactic measures of stars only a hundred light-years away.

Sounding Space

This by no means exhausts the list of modern methods of surveying space. Any standard type of astronomical object, which is easily recognisable and emits the same amount of light no matter where it occurs, provides an obvious means of measuring astronomical distances, for when once the intrinsic luminosity of such an object has been determined, the distance of every example of it can be estimated from its apparent brightness.

Cepheid variables of assigned periods provide the most striking instance of such standard objects, but three others are available, although they are not so generally useful as Cepheids. First comes another type of variable star, the "long-period variables" already mentioned, which are generally similar to Cepheids except that their light fluctuates much more slowly. These stars are intrinsically far more luminous even than Cepheids, many of them being 10,000 times as luminous as the sun. They are accordingly visible at enormous distances, and may ultimately be found to provide a means of sounding depths of space at which even Cepheids are lost to sight.

Next come "novae" or new stars. Every now and then an ordinary star in the sky suddenly bursts out in a phenomenal blaze of light, shining with perhaps a thousand times its original brilliance. The cause of these violent outbursts is still a matter for debate, and no thoroughly convincing explanation has as yet been given. A study of comparatively near novae has, however, provided information as to the luminosity of the average nova when at its brightest, and as novae appear in various parts of the sky, and particularly in the extra-galactic nebulae, they provide a rough means of measuring stellar and nebular distances.

Blue stars provide yet another method. These are exceedingly luminous, and they vary but little in intrinsic luminosity. Moreover, the luminosity of any particular star can generally be estimated fairly closely from the quality of the light it emits, by methods which will be explained later. This makes it possible to determine the distances of blue stars, and so of course of the astronomical objects in which they occur.

Still two other methods of a different kind may be briefly mentioned. Dr. W. S. Adams, Director of Mount Wilson Observatory, and others have found that certain definite peculiarities in the spectra of certain classes of stars convey information as to the intrinsic brightness of the star emitting them; with this information it is easy to estimate the star's distance from its apparent brightness. This is commonly described as the method of Spectroscopic Parallaxes.

Finally the diffuse cloud of nebular matter which is spread through interstellar space (p. 29) is found to affect the quality of light travelling through it, so that a star's spectrum gives an indication of the amount of cloud through which the light of the star has travelled, and this again provides a rough means of estimating distances inside the galactic system.

Globular Clusters. The law of Cepheid luminosity was first used by Hertzsprung to estimate the distance of the Lesser Magellanic Cloud, the study of which had been responsible for the original discovery of the law. Shapley subsequently used it to determine the distances of the rather mysterious groups of stars known as "Globular Clusters." A typical example of these is shewn in Plate IX. About 100 of these clusters are known and they all look pretty much alike, except for differences in apparent size. Even these latter can be traced mainly to differences of distance, so that the globular clusters are probably almost identical objects, and Plate IX might almost be regarded as a picture of any one of them. Cepheid variables abound in them all.

Shapley found the nearest globular cluster, ω Centauri, to lie at a distance of about 22,000 light-years, the furthest, N.G.C. 7006, being about ten times as remote, at a distance of 220,000 light-years. At such distances the parallactic method of measuring distances would of course fail hopelessly. The parallactic orbit of a star at 220,000 light-years'

PLATE IX

Mt. Wilson Observatory.

The Globular Cluster (*M* 13) in Hercules.

distance is about the size of a pin-head held at a distance of 4000 miles; no telescope on earth could detect, still less measure, such an orbit.

The mere figure of 220,000 light-years can convey but little conception of the distance of this remotest of star-clusters from us. We may apprehend it better if we reflect that the light by which we see the cluster started on its long journey from it to us somewhere about the time when

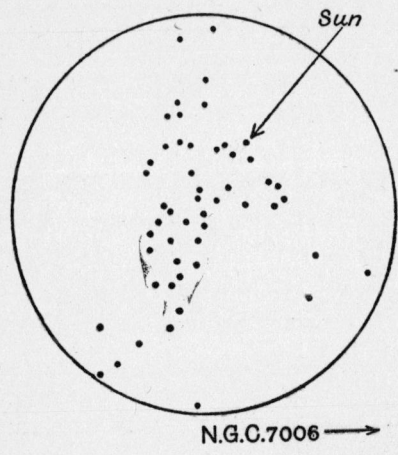

Fig. 8.—The arrangement of the Globular Clusters.

primaeval man first appeared on earth. Through the childhood, youth and age of countless generations of men, through the long prehistoric ages, through the slow dawn of civilisation and through the whole span of time which history records, through the rise and fall of dynasties and empires, this light has travelled steadily on its course, covering 186,000 miles every second, and is only just reaching us now. And yet this enormous stretch of space does not carry us to the confines of the universe; we shall now see that in all probability it has barely carried us to the confines of the galactic system.

Shapley has mapped out the complete system of the globular clusters, and finds that they occupy an oblong region, lying on both sides of the plane of the Milky Way, its greatest diameter of about 250,000 light-years lying in this plane, and its two transverse diameters being considerably shorter. The sun is nearer to the edge of this oblong region than to its centre, which explains why all the globular clusters appear in one half of the sky, as Hinks first noted in 1911. The general arrangement is shewn in fig. 8. The page of the book represents the plane of the Milk Way, the various dots representing the points in this plane which are nearest to the different clusters, so that the diagram exhibits the system of globular clusters as they would appear to an observer out in space who viewed the galactic plane "full-on." All the globular clusters except N.G.C. 7006 lie within a circle of about 125,000 light-years' radius, having its centre at about 50,000 light-years from the sun.

The Arrangement of the Galactic System. Although the matter has long been one of vigorous controversy, it is now becoming clear that the region of space mapped out by these globular clusters approximately coincides with that occupied by the galactic system itself. Herschel and Kapteyn appear to have been in error in supposing the centre of the galactic system to be in the neighbourhood of the sun; Shapley believes that it lies somewhere in a massive star cloud in the constellations of Scorpio and Ophiuchus at a distance of about 47,000 light-years from the sun. There is what Shapley describes as a "local system" of fairly bright stars surrounding the sun, and the error of identifying this with the main galactic system has apparently been responsible for

PLATE X

The Region of ϱ Ophiuchi.

E. E. Barnard.

a large part of the confusion which has hitherto beset the problem of the architecture of the galaxy. This local system has the same flattened shape as the main system, but it does not lie exactly in the plane of the Milky Way, being inclined at an angle of about 12 degrees to it. Fig 9 shews a cross-section of the system, as it is now imagined to lie.

Various attempts have been made to estimate the total number of stars in the galactic system. The whole question is rather in the melting-pot at the moment, all the early estimates being invalidated by the confusion between the main and the local system which has hitherto prevailed. Seares estimated the total number at somewhere about 30,000

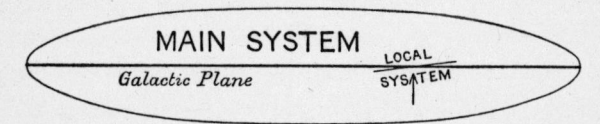

FIG. 9.—Diagrammatic scheme of cross-section of the Galactic System. The sun is at the head of the arrow.

million, but the estimate was largely conjectural. Even photographically, the giant 100-inch telescope at Mount Wilson only shews about 1500 million stars in all. A great number of these are so faint as to be at the extreme limit of vision, whence it is clear that a slight increase in telescopic power would bring a great many more stars into visibility, and that the total number must quite certainly be far above 1500 million. Seares' estimate of 30,000 million was based on a careful extrapolation from observed data. More recently Shapley has spoken of 100,000 million as the possible round number of stars in the galactic system.

Another way of estimating the total number of stars in the galactic system is that of weighing them *en masse*. Individual stars far away from the centre of the galactic system must be describing orbits under the gravitational pull of the system as a whole; it is this pull which prevents the stars from scattering away into space, and so keeps the galactic system in being. From the estimated orbital speeds of the stars in the vicinity of the Sun, Eddington has calculated that the stars inside the sun's orbit must have a total weight about equal to that of 270,000 million suns. This would seem to involve that the total number of stars in the system must be well over 300,000 million.

Again we are confronted with the difficulty of visualising such large numbers. With perfect eyesight on a clear moonless night we can see about 3000 stars. Imagine each of these 3000 stars to spread out into a complete sky-full of 3000 new stars, and we are contemplating 9 million stars, which is still only the number visible in a telescope of 5 inches aperture. We probably cannot ask our imagination to play the same trick for us a second time, but if it can be persuaded to do so, and if we can think of each of these 9 million stars as again generating a whole sky-full of stars, we still have only 27,000 million stars within our purview—a number which is less than Seares' estimate of the total number of stars in the galactic system and far below that of Eddington. Or, if we prefer, let us notice that the number of stars photographically visible in the 100-inch telescope, namely 1500 million, is about equal to the number of men, women and children in the world. Each inhabitant of the earth—each man, woman and child living in the five continents or travelling on the seven seas—can be allowed to choose his own particular star, and can then repeat the process from 20 times (Seares) to 200 times (Eddington) without going outside the galactic system.

After this we can still go exploring outside the galactic

system and find more and ever more stars. The galactic system, with its 30,000 million or more of stars, no more contains all the stars in space than one house contains all the inhabitants of Great Britain. There are millions of other houses and millions of other families of stars.

The Extra-Galactic Nebulae. We have already spoken of the faint nebulous objects which Herschel described, somewhat conjecturally, as "island universes." These are the other houses in which other families of stars are to be found. The most powerful of modern telescopes shew that they consist, in part at least, of huge clouds of stars. Just as a powerful microscope shews that a puff of cigarette smoke, in spite of its appearance of continuity, consists of a cloud of minute but quite distinct particles, so a powerful modern telescope breaks up the light from the outer regions of these nebulae into distinct spots of light; the nebula is resolved into a cloud of shining particles, just as the Milky Way was in Galileo's tiny telescope of three centuries ago. Plate XI shews an example; it represents a magnification of a small area in the top left-hand corner of the Great Nebula in Andromeda (M 31) already shewn in Plate IV (p. 30), and the resolution into distinct spots of light is unmistakable. We know that some at least of these spots of light are stars, because we recognise them as Cepheid variables, their light shewing the unmistakable characteristic fluctuations of the familiar Cepheid variables nearer home. The other shining particles are of comparable brightness and shew about the range of brightness above and below that of the Cepheids which is needed to justify us in supposing that they are ordinary stars.

From the observed periods of fluctuation of their Cepheid variables, in combination with the other methods just explained, Dr. Hubble of Mount Wilson Observatory has recently found that even the nearest of these nebulae, namely the nebula M 33 shewn in Plate XX (p. 201), is so remote that light takes some 850,000 years to travel from it to us. The Great Nebula M 31 in Andromeda (p. 29) is at the slightly greater distance of about 900,000 light-years. This abundantly proves that these nebulae lie right outside the galactic system, justifying the term "extra-galactic" nebulae.

One might attempt to estimate the total number of stars in these nebulae by counting those visible in a selected average small area, but more precise methods are available. Just as we have supposed that the outermost stars in the galactic system are describing orbits under the gravitational attraction of the galaxy as a whole, so we must suppose that the outermost stars in a nebula are describing orbits under the gravitational attraction of the main mass of the nebula; the forces which keep them from running away from the nebula are similar to those which keep the earth moving in its orbit round the sun. If so, we can weigh the nebulae, precisely in the same way as we weigh the sun. Dr. Hubble in this way estimates that the weight of the Great Nebula in Andromeda (M 31) shewn in Plate IV, must be about 3500 million times that of the sun, while the nebula N.G.C. 4594 in Virgo shewn in Plate XV (p. 194), must have about 2000 million times the weight of the sun. In general it seems likely that each of the extra-galactic nebulae contains about enough matter to make some 2000 million stars.

This is not the same thing as saying that each nebula already contains 2000 million stars. While many of these nebulae appear to consist largely of clouds of stars, yet most of them contain also a large central region which no telescopic power has so far succeeded in resolving into distinct points. For instance, Plate XII shews the central region of

PLATE XI

Magnification of a part (left-hand top corner) of the Great Nebula in Andromeda (M 31), which is shewn complete in Plate IV (p. 30).

Mt. Wilson Observatory.

PLATE XII

Mt. Wilson Observatory.

Magnification of the central region of the Great Nebula in Andromeda (*M* 31).

the Great Nebula in Andromeda magnified to the same degree as the left-hand top corner shewn in Plate XI, and this is clearly not resolved into stars in the same way as the outer regions shewn in Plate XI. The whole of the nebula N.G.C. 4594 in Virgo, shewn in Plate XV, also refuses to be resolved into separate stars. We shall find reasons later (Chapter IV) for interpreting these central regions as masses of gas which are destined in time to form stars, but have not yet done so. We shall in fact find that the nebulae are the birthplaces of the stars, so that each nebula consists of stars born and stars not yet born. It is the total weight of stars already born and of matter which is destined to form stars that aggregates 2000 million suns.

About 2,000,000 of these extra-galactic nebulae are visible in the great 100-inch telescope. They appear to be scattered with a tolerable approach to uniformity through space, their average distance apart being something of the order of 2,000,000 light-years. The most distant of them is about 140 million light-years from us.

The Remotest Depths of Space. This is the greatest distance which the human eye has so far seen into space. The 220,000 light-years which formed the diameter of the galactic system seemed staggeringly large at first, but we are now speaking of distances some 600 times greater. For all but a 500th part of its long journey, the light by which we see this remotest of visible nebulae travelled towards an earth uninhabited by man. Just as it was about to arrive, life sprang into being on earth, and built telescopes to receive it. So at least it appears when viewed on the astronomical scale. Yet even this last 500th part of the journey covers the lives of 10,000 generations of men, through all of which, as well as through 500 times as great a span of time, the light has been travelling steadily onward at 186,000 miles a second.

There are so many faint nebulae at the very limit of vision of the 100-inch telescope, that it seems certain that a still larger telescope would reveal a great many more. The 200-inch telescope, which it is hoped will shortly be built, having twice the aperture of the present 100-inch, ought to probe twice as far into space, and so may perhaps be expected to shew about eight times as many, or 16 million, nebulae.

The Structure of the Universe

So far every increase of telescopic power has carried us deeper and deeper into space, and space has seemed to expand at an ever-increasing rate. We may well ask whether this expansion is destined to go on for ever: are there any limits at all to the extent of space?

Even a generation ago, I think most scientists would have answered this last question in the negative. They would have argued that space could be limited only by the presence of something which is not space. We, or rather our imaginations, could only be prevented from journeying for ever through space by running up against a wall of something different from space. And, hard though it may be to imagine space extending for ever, it is far harder to imagine a barrier of something different from space which could prevent our imaginations from passing into further space beyond.

The argument is not a sound one. For instance, the earth's surface is of limited extent, but there is no barrier which prevents us from travelling on and on as far as we please. A traveller who did not understand that the earth's surface is spherical, would naturally expect that longer and longer journeys from home would for ever open up new tracts of country awaiting exploration. Yet, as we know, he would necessarily be reduced in time to repeating his own tracks. As a result of its curvature, the earth's surface, although unlimited, is finite in extent.

The Theory of Relativity

Through his theory of relativity, Einstein claims to have established that space also, although unlimited, is finite in extent. The total volume of space in the universe is of finite amount just as the surface of the earth is of finite amount, and for the same reason; both bend back on themselves and close up. The analogy is valid and useful only so long as we are careful to compare the whole of space to the surface of the earth, and not to its volume. The volume of the earth is also finite in amount, but for quite different reasons. A mole which burrowed on and on through the earth in a straight line would come in time to something which is not earth—it would emerge into the open air; but we can go on and on over the surface of the earth without ever coming to anything which is not the surface of the earth. The properties of space are those of the surface, not of the volume, of the earth.

As a consequence of space bending back into itself, a projectile or a ray of light can travel on for ever without going outside space into something which is not space, and yet it cannot go on for ever without repeating its own tracks. For this reason it is probable that light can travel round the whole of space and return to its starting point, so that if we pointed a sufficiently powerful telescope in the right direction in the midnight sky, we should see the sun and its neighbours in space by light which had made the circuit of the universe. We should not see them as they now are, but as they were many millions of years ago. Light which had left the sun so long ago would have travelled round almost the whole of space and then, just as it was about to complete the circle, it would be caught in our telescope instead of being allowed to start on its second journey round space.

This curvature of space has other functions than that which it performs on the grand scale, of limiting the total volume of space. Before Einstein's day the curvatures of the paths of planets, cricket balls and projectiles in general were all attributed to the pull of a "force" of gravitation. The theory of relativity dismisses this supposed force as a pure illusion, and attributes the curved paths of projectiles of all kinds to their efforts to keep a straight track through a curved space. This curved space is not, it is true, the ordinary space of the astronomer. It is a purely mathematical and probably wholly fictitious space, in which the astronomers' space and the astronomers' time are inextricably bound together and enter as equal partners. To be absolutely exact, there are four equal partners. The first three are the three dimensions of ordinary space—breadth, width, and height, or, if we prefer, north-south, east-west and up-down. The fourth is ordinary time measured in a way appropriate to the way in which we have measured our space (a year of time corresponding to a light-year of space, and so on), and then multiplied by the square-root of -1. This last multiplication by the square-root of -1 is of course the remarkable feature of the whole affair. For the square-root of -1 has no real existence; it is what the mathematician describes as an "imaginary" number. No real number can be multiplied by itself and give -1 as the product. Yet it is only when time is measured in terms of an imaginary unit of $\sqrt{(-1)}$ years that there is true equal partnership between space and time. This shews that the equal partnership is purely formal—it is nothing but a convenient fiction of the mathematician. Indeed had it been anything more, our intuitive conviction that time is something essentially different from space could have had no basis in experience and so would have vanished ere now.

These complications with respect to time need not concern us here; the essential point is that Einstein's theory of relativity teaches that space ultimately bends back on itself like the earth's surface, so that the total amount of space is finite.

The Cosmology of Einstein. According to Einstein's original theory, the dimensions of space are determined by the amount of matter it contains. The more matter there is, the smaller space must be, and conversely; space could only be of literally infinite extent if it contained no matter at all. The problem of determining the extent of space accordingly reduces to that of determining how much matter it contains. We have no means of estimating how much matter may exist outside those regions of space which are within the reach of our telescopes, but within these regions matter seems to be fairly uniformly distributed in the form of extra-galactic nebulae.

From the known weights of these, Hubble estimates that the mean density of matter in space must be about 1.5×10^{-31} times that of water. On the assumption that matter is distributed with this density through the whole of space, including those parts which our telescopes have not yet penetrated, we can calculate quite definitely that the radius of space is 84,000 million light-years, or 600 times the distance of the furthest visible nebula. The journey round space would take 500,000 million light-years, and if ever our telescopes shew us the solar system from behind, we shall see it as it was 500,000 million years ago.

Thus, according to Einstein's original theory, even the 140 million light-years through which we can range with our telescopes form only a small fraction of the whole of space —something like one part in a thousand million. There is plenty of space still awaiting exploration. It is perhaps not surprising. Mankind, who has been possessed of telescopes for only 300 years out of the 300,000 of his residence on earth, could hardly hope to discover the whole of space in so short a time. Our astronomer explorers are moving from island to island in the small archipelago which surrounds their home in space, but they are still far from circumnavigating the globe. And, just as the earliest geographers tried to estimate the size of the earth, long before they thought of circumnavigating it, from the curvature of a small part of its surface, so astronomers are now trying to form estimates, although necessarily vague, of the size of the whole universe from the curvature of that part of it with which they are already acquainted.

The general theory of relativity has long passed the stage of being regarded as an interesting speculation. It not only accounts for phenomena of planetary motion before which Newton's law of gravitation failed, but it has predicted other phenomena—the apparent displacements of stars near the sun at an eclipse, resulting from the light by which we see them being bent as it passes through the sun's gravitational field, and a certain displacement of stellar spectra towards the red end—which were entirely unsuspected when the predictions were first made, but have subsequently been fully confirmed by observation. Indeed the theory has by now qualified as one of the ordinary working tools of astronomy. It has been used to measure the diameter of the small faint star Sirius B, the companion to Sirius (p. 246), as well as to test the nature of the stars at the centres of the "planetary nebulae" (p. 303).

Nevertheless, the general theory of relativity does not lead up to Einstein's cosmology in a unique way. It is perfectly possible for the former to be true and the latter false. The general theory of relativity fixes the attributes of any small fraction of the universe quite definitely, but leaves open several alternative ways in which these small fractions can be pieced together to form a whole. Einstein's particular view of the cosmos cannot therefore claim the prestige which attaches to the general theory of relativity as a whole. And indeed it has recently fallen somewhat into disfavour, and appears likely to be superseded by an alternative cosmology which de Sitter of Leiden propounded and developed in some detail in 1917.

The Cosmology of de Sitter. Let us first try to understand the essential differences between these two cosmologies.

Einstein's cosmology supposes that the size of the cosmos is determined by the amount of matter it contains. If it was decided, at the creation, to create a universe containing a certain amount of matter which was to obey certain natural laws, then space must at once have adjusted itself to the size suited for containing just this amount of matter and no more. Or, if the size of the universe and the natural laws were decided upon, the creation of a certain definite amount of matter became an inevitable necessity. De Sitter's universe is less simple, or, if we prefer so to put it, allows more freedom of choice in its creation. After the laws of nature had been fixed, it was still possible to make a universe of any size, and to put any amount of matter, within limits, into it. Looked at from the strictly scientific point of view, Einstein's universe has one element of arbitrariness fewer than de Sitter's universe, and to this extent it has the advantage of simplicity.

On the other hand this simplicity is acquired at a price. The fundamental corner-stone of the whole theory of relativity is the equal partnership of space and time in the sense already explained. Einstein's cosmology gains its simplicity only at the expense of supposing that this equality of partnership disappears when we view the cosmos as a whole. It supposes that space and time are indistinguishable (in the purely formal sense already indicated) only to a being whose experience is limited to a small fraction of the universe; they become utterly distinct for a being who can range through the whole of space and time. It is not altogether clear how much weight ought to be attached to this objection, if objection it is. Real space and real time undoubtedly are distinct. Even if we deny the reality of both, they still remain distinguishable as modes of perception. What reproach, then, can it be to a cosmology that it admits that, in the last resort, when the universe is contemplated on the grand scale, space and time resolve themselves into distinct types of entity? Somehow we knew it already, before ever we began to contemplate the universe on the grand scale.

Whatever the answer to this last question may be, de Sitter's cosmology avoids all possible reproach by maintaining the equal partnership of space and time, not only in individual fractions of the cosmos, but throughout the cosmos as a whole. It will of course be understood that we are still speaking of equal partnership in the purely formal sense already explained, a light-year entering the cosmology on the same footing as the square-root of -1 years. Even de Sitter's cosmology does not pretend that a light-year (9.46 million million kilometres) is the same thing as twelve months.

Although Einstein's main theory of relativity has been amply confirmed by observation, the cosmological part of it did not predict any special features such as permitted of a direct observational test. De Sitter's cosmology, on the other hand, predicts that the spectra of all distant objects must shew a displacement towards the red, of amount depending on the distance of the object. The equal partnership of space and time results in the vibrations of the light-waves emitted by any specified source being slower in distant than in near parts of the universe; the stream of time rolls more rapidly just where we happen to be than anywhere else. This sounds paradoxical at first, but examination shews that it is not; de Sitter is not asking us to return to a geocentric universe, because he shews that the inhabitant of a distant star would also find that terrestrial atoms were keeping slower time than his own. The paradox is completely resolved by the concept of the relativity of all measures of space and time.

This displacement to the red as a result of mere distance is peculiar to de Sitter's cosmology. It is additional to the displacement which, as all cosmologies agree, the spectrum of a moving body must shew as the result of its motion, this latter being towards the red only if the body is receding from the earth (p. 49). On de Sitter's cosmology, the two displacements are not entirely independent, for it is an essential feature of this cosmology that near bodies should tend to move further apart from one another. Just as bits of straw thrown together into a stream tend to get separated as they float down the stream, so objects in de Sitter's universe move further apart as they float down the stream of time.

Thus on de Sitter's theory a displacement of spectral lines to the red cannot be interpreted as evidence either of motion or of distance; it is a mixture of both. This does not mean that we have been altogether wrong in deducing the velocities of stars in the galactic system from the observed displacements of their spectral lines. No appreciable displacement is produced by distance alone, unless this distance forms an appreciable fraction of the radius of the universe. Systematic displacements to the red are, it is true, observed in the spectra of the most distant stars, but they are of very small amount. It is only when we look to the remote extra-galactic nebulae that we can expect to observe the effect in appreciable strength.

Now it has long been one of the outstanding puzzles of astronomy that the spectra of the distant nebulae are uniformly displaced towards the red. The observed displacements are not small. Interpreted as velocities, many of them would represent speeds of over 1000 miles a second, while the spectrum of the faint nebula N.G.C. 7619, which is probably at a distance of over 20 million light-years, shews a displacement corresponding to a speed of 2350 miles a second. If de Sitter's theory is rejected, almost all the extra-galactic nebulae must be running away from us with terrific, almost unimaginable, speeds. Yet we can hardly reintroduce simplicity by adopting de Sitter's theory, and treating the whole apparent stampede of nebulae as spurious, since this theory involves that the nebulae may well, in actual fact, be running away from us, scattering being an inherent property of objects in a de Sitter universe.

The fact that the spectra of the most distant nebulae shew

these large displacements provides a certain presumption in favour of the truth of de Sitter's cosmology; this at least explains them twice over, while no other theory can explain them at all. If we tentatively accept this cosmology, then each observed spectral shift must be regarded as the sum of two parts, one arising in the ordinary way from a recession of the nebula, and the other arising merely from its distance.

Imagine that at the beginning of time the nebulae were much nearer to one another than they now are, and that they have merely obeyed the inherent tendency to scatter implied in de Sitter's cosmology—or rather the tendency of the flowing stream of time to scatter them. At any subsequent time the most distant nebulae would be receding most rapidly, and it can be shewn that their speeds of scattering would be exactly proportional to their distances from us. A general preliminary study by Dr. Hubble shews that on the whole the spectral displacements are largest for the most distant nebulae, and that their amounts are roughly proportional to the distances of the nebulae from us. If we interpret the whole of the observed displacements purely as evidence of recession, we can calculate that the radius of the universe is about 2000 million light-years, or some fourteen times the distance of the furthest visible nebula. With so large a radius of the universe, the further displacement resulting from the mere distance of even remote nebulae is negligible, so that our assumption that the displacements arise almost entirely from velocities of recession receives *à posteriori* vindication. If the observed displacements of the nebular spectra had been strictly proportional to their distances from us, we should have obtained a consistent explanation of the observed facts by assuming that we lived in a de Sitter universe having a radius of about 2000 million light-years.

This provides a simple and rather fascinating picture of the universe, but there are many reasons against supposing that it is a true one. In the first place, if we interpret the spectral displacements as evidence of velocity alone, the speeds of the nebulae are very far from being (as the foregoing picture would require) accurately proportional to their distances from us. The Magellanic Clouds, at distances of only about 100,000 light-years shew velocities of recession of about 150 miles a second, which is about seven times too large, while the two nearest nebulae, at a distance of about a million light-years, shew *velocities of approach* of about 200 miles a second. Whatever interpretation we put on the spectral displacements, these two nebulae must be coming towards us with terrific speeds. Again, if the observed displacements represent mere scattering, we can calculate the time since this scattering began; it proves to be many thousands of millions of years. Enormous though such a length of time is, it does not appear to be enough. We shall see later (Chapter III) how time leaves its mark, its wrinkles and its grey hairs, on the stars, so that we can guess their ages tolerably well, and the evidence is all in favour of stellar lives, not of thousands of millions, but of millions of millions, of years. If the nebulae owe their present motions to mere scattering, then the stars must have lived the greater parts of their lives before this scattering began. Such a hypothesis seems too artificial for acceptance, at any rate so long as any alternative is open.

Of course we must frankly admit that our estimates of stellar ages may be found to need revision. Indeed they have been calculated on the supposition that no appreciable scattering of the type required by de Sitter's cosmology has ever taken place. If not only the nebulae, but also the stars composing the galactic system, were huddled together at the beginning of time, our estimates of the lives of the stars would have to be substantially shortened, and it is conceivable, although I think very unlikely, that they could be reduced to lengths of the kind we have just considered. In this event we might still try to obtain a consistent picture by supposing that the main masses of the universe came into being in a comparatively small region of space some thousands of millions of years ago and had been scattering ever since. Such a supposition, however, fails entirely to account for the rapid speeds of approach of the nearest nebulae, and its adoption renders a large part of modern astronomy meaningless, so that we shall not discuss it further at present.

Before we part from de Sitter's cosmology, let us tentatively examine the effect of flying to the other extreme, and supposing that the spectral displacements are caused predominantly by distance rather than speed. We are not of course free to suppose that the nebulae have no motions at all in space. Distance can only produce displacements to the red, and many nebular spectra shew displacements to the violet which can only be produced by rapid motions of approach. The Magellanic Clouds and the nearest of the nebulae are so near that their distances can hardly contribute much to the observed displacements, so that these must originate in true velocities, some of approach and some of recession, averaging some 175 miles a second or so. We may properly suppose that all the nebulae now have, and have always had, random velocities of this order. When space is filled with nebulae moving in this random way, the tendency to scattering disappears automatically—companions tend to become separated in the crowd, but this is not the same thing as saying that the crowd tends of itself to become sparser. After allowing for these random motions, residual displacements of the nebular spectra remain, and these are found to be greatest for the most distant nebulae. If we now interpret these residual displacements as arising from distance only, we can calculate that the radius of the universe must be something like 80 million light-years, or little more than half the distance of the farthest visible nebula. The journey of light round space and back to the starting point would take about 500 million years, so that if we could see even three or four times as far into space as we now see, the nebulae nearest to the sun ought to be visible by light which had travelled the long way round the universe.

These are not the mere irresponsible reveries of a heated imagination. It has been quite seriously suggested that two faint nebulae (h 3433 and M 83) may actually be our two nearest neighbours in the sky, M 33 and M 31, seen the long way round space. If so, we see the fronts of two objects when we look at M 33 and M 31, and the backs of the same two objects when we point our telescopes in exactly the opposite directions and look at h 3433 and M 83. No doubt this is only a conjecture, and perhaps rather a wild one, but many more startling conjectures have been made in astronomy, and subsequently proved to be true.

In de Sitter's original form of this cosmology, light would take an infinite time to travel round the universe, and this would prevent any object being seen by light which had travelled the long way round. This results from de Sitter having considered only the ideal case of a universe entirely empty of all matter. With even a little matter in the universe, the path of a ray of light would presumably bend back on itself and return to its starting point after a finite time.

A more serious difficulty arises from the circumstance that

a body does not move with a constant speed through the de Sitter universe, even when it is acted on by no forces. Mere motion through space induces changes of velocity, and the smaller the radius of the universe, the greater these changes of velocity must be. If the radius of the universe were only the 80 million light-years we have just calculated, these changes of velocity would be so great that it becomes difficult to see why the nebular velocities are not *greater* than they actually are.

If the universe is really built according to the de Sitter cosmology, the truth seems likely to lie somewhere between the two extremes we have just considered, and, in all probability, nearer to the former than the latter. We may think of the radius of space as hundreds of millions of light-years at least, and of the journey of light round the whole of space as occupying thousands of millions of years.

All these discussions of the structure of space are of course highly speculative, but they agree in suggesting the general conclusion that, if we cannot yet see the whole of space, we can at least survey a comparatively large fraction of it. Our astronomer-explorers may not as yet have circumnavigated the globe, but they are perhaps discovering America, and we can well imagine that even the next generation will have completed the circumnavigation of space, and will think of a finite but unbounded space in the same way, and with the same ease, as we think of the finite but unbounded surface of the earth.

A Model of the Universe

We found it difficult enough to visualise the $4\frac{1}{4}$ light-years which constitute the distance to the nearest star, so we may be well advised not even to attempt to visualise this last distance of thousands of millions of light-years, the conjectured circumference of the universe. Yet we may try to see all these distances in proper proportion relative to one another by the help of a model drawn to scale. We can escape the effort of trying to imagine unimaginably great distances by keeping the scale very small.

The earth, travelling 1200 times faster than an express train, makes a journey of 600 million miles around the sun every year. Let us represent this journey by a pin-head 1/16 of an inch in diameter. This fixes the scale of our model; the sun has shrunk to a minute speck of dust 1/3400 inch in diameter, while the earth is a still more minute speck which is too small to be seen at all even in the most powerful of microscopes. On this scale the nearest star in the sky, Proxima Centauri, must be placed about 225 yards away, and to contain even the hundred stars nearest to our sun in space, the model must be a mile high, a mile long and a mile wide.

Let us go on building the model. We may think of stars indiscriminately as specks of dust, because their sizes vary about as much as the sizes of specks of dust. In the vicinity of the sun we must place specks of dust at average distances of about a quarter of a mile apart. In other regions of space they are generally even farther apart, for, owing to the presence of the "local cluster," the immediate neighbourhood of the sun happens to be a rather crowded part of the sky. We go on building the model for hundreds of miles in every direction, and then, if we are building in a direction well away from the galactic plane, the specks of dust begin to thin out; we are approaching the confines of the galaxy. In the galactic plane itself we build out for about 7000 miles before we come to the farthest globular cluster, and still we are inside the galactic system. With our earth's long yearly journey round the sun as a pin-head the whole galactic system is about the size of the American continent. It may be well to pause and try to visualise the relative sizes of a pin-head and of the American continent, before we go on with our mental model-building.

After we have finished the galactic system, we must travel about 30,000 miles before we begin to set up the next bit of our model, at any rate if we are keeping it to scale. At this distance we place the next family of stars, a family which is probably substantially smaller and more compact than our own galactic family, but is comparable with it both in size and in numbers. So we go on building our model—a family of thousands of millions of stars every 30,000 miles or so—until we have two million such families. The model now stretches for about four million miles in every direction. This represents as far as we can see into space with a telescope; we can imagine the model going on, although we know not how nor where—all we know is that the part so far built represents only a fraction of the universe.

Every galactic system or island universe or extra-galactic nebula contains thousands of millions of stars, or gaseous matter destined ultimately to form thousands of millions of stars, and we know of two million such systems. There are, then, thousands of millions of millions of stars within the range covered by the 100-inch telescope, and this number must be further multiplied to allow for the parts of the universe which are still unexplored. At a moderate computation, the total number of stars in the universe must be something like the total number of specks of dust in London. Think of the sun as something less than a single speck of dust in a vast city, of the earth as less than a millionth part of such a speck of dust, and we have perhaps as vivid a picture as the mind can really grasp of the relation of our home in space to the rest of the universe.

An alternative procedure would have been to construct our scale-model by taking all the specks of dust in London and spreading them out to the right distances to represent the various stars in space. The average actual distances between specks of dust in London is a quite small fraction of an inch; to get our model to correct scale, this distance must be increased to about a quarter of a mile, even when we are building the part which represents the crowded part of space round the sun. If we build our model in this way, we obtain a vivid picture of the emptiness of space. Empty Waterloo Station of everything except six specks of dust, and it is still far more crowded with dust than space is with stars. This is true even of the comparatively crowded region inside the galactic system; it takes no account of the immense empty stretches between one system of stars and the next. On averaging throughout the whole of the model, the mean distance of a speck of dust from its nearest neighbour proves to be something like 80 miles. The universe consists in the main not of stars but of desolate emptiness—inconceivably vast stretches of desert space in which the presence of a star is a rare and exceptional event.

Let us in imagination take up a position in space somewhere near the sun, and watch the stars moving past with speeds about 1000 times that of an express train. If space were really crowded with stars our position would be as unenviable as if we sat down in the middle of Regent Street to watch the the traffic go by—our life though thrilling would be brief. Yet, as exact calculation shews, the stellar traffic is so little crowded that we would have to wait about a million million million years before a star ran into us. Put in another form, the calculation shews that any one star may expect to move for something of the order of a million million million years before colliding with a second star. The stars move blindly through space, and the players in the stellar blind-man's-buff are so few and far between that the chance of encountering another star is almost negligible. We shall see later that this concept is of the profoundest significance in our interpretation of the universe.

CHAPTER II

Exploring the Atom

So far our exploration of the universe has been in the direction from man to bigger things than man; we have been exploring ranges of space which dwarf man and his home in space into utter insignificance. Yet we have explored only about half the total range of the universe; an almost equal range awaits exploration in the direction of the infinitely small. We appreciate only half of the infinite richness of the world around us until we extend our survey down to the smallest units of matter. This survey has been first the task, and now the brilliant achievement, of modern physics.

It may perhaps be asked why an account of modern astronomy should concern itself with this other end of the universe. The answer is that the stars are something more than huge inert masses; they are machines in action, generating and emitting the radiation by which we see them. We shall best understand their mechanism by studying the ways in which radiation is generated and emitted on earth, and this takes us right into the heart of modern atomic physics. In the present book we naturally cannot attempt to cover the whole of this new field of knowledge; we shall concern ourselves only with those parts which are important for the interpretation of astronomical results.

ATOMIC THEORY

As far back as the fifth century before Christ, Greek philosophy was greatly exercised by the question of whether in the last resort the ultimate substance of the universe was continuous or discontinuous. We stand on the sea-shore, and all around us see stretches of sand which appear at first to be continuous in structure, but which a closer examination shews to consist of separate hard particles or grains. In front rolls the ocean, which also appears at first to be continuous in structure, and this we find we cannot divide into grains or particles no matter how we try. We can divide it into drops, but then each drop can be subdivided into smaller drops, and there seems to be no reason, on the face of things, why this process of subdivision should not be continued for ever. The question which agitated the Greek philosophers was, in effect, whether the water of the ocean or the sand of the seashore gave the truest picture of the ultimate structure of the substance of the universe.

The school of Democritus, Leucippus and Lucretius believed in the ultimate discontinuity of matter; they taught that any substance, after it had been subdivided a sufficient number of times, would be found to consist of hard discrete particles which did not admit of further subdivision. For them the sand gave a better picture of ultimate structure than the water, because, or so they thought, sufficient subdivision would cause the water to display the granular properties of sand. And this intuitional conjecture is amply confirmed by modern science.

The question is, in effect, settled as soon as a thin layer of a substance is found to shew qualities essentially different from those of a slightly thicker layer. A layer of yellow sand swept uniformly over a red floor will make the whole floor appear yellow if there is enough sand to make a layer at least one grain thick. If, however, there is only half this much sand, the redness of the floor inevitably shews through;

it is impossible to spread sand in a uniform layer only half a grain thick. This sudden change in the properties of a layer of sand is of course a consequence of the granular structure of sand.

Similar changes are found to occur in the properties of thin layers of liquid. A teaspoonful of soup will cover the bottom of a soup plate, but a single drop of soup will only make an untidy splash. In some cases it is possible to measure the exact thickness of layer at which the properties of liquids begin to change. In 1890 Lord Rayleigh found that thin films of olive oil floating on water changed their properties entirely as soon as the thickness of the film was reduced to below a millionth of a millimetre (or a 25,000,000th part of an inch). The obvious interpretation, which is confirmed in innumerable ways, is that olive oil consists of discrete particles—analogous to the "grains" in a pile of sand—each having a diameter somewhere in the neighbourhood of a 25,000,000th part of an inch.

Every substance consists of such "grains." They are called molecules, and the familiar properties of matter are those of layers many molecules thick; the properties of layers less than a single molecule thick are known only to the physicist in his laboratory.

Molecules

How are we to break up a piece of substance into its ultimate grains, or molecules? It is easy for the scientist to say that, by subdividing water for long enough, we shall come to grains which cannot be subdivided any further; the plain man would like to see it done.

Fortunately the process is one of extreme simplicity. Take a glass of water, apply gentle heat underneath, and the water begins to evaporate. What does this mean? It means that the water is being broken up into its separate ultimate grains or molecules. If the glass of water could be placed on a sufficiently sensitive spring balance, we should see that the process of evaporation does not proceed continuously, layer after layer, but jerkily, molecule by molecule. We should find the weight of the water changing by jumps, each jump representing the weight of a single molecule. The glass may contain any integral number of molecules but never fractional numbers—if the fractions of a molecule exist, at any rate they do not come into play in the evaporation of a glass of water.

The Gaseous State. The molecules which break loose from the surface of the water as it evaporates form a gas—water-vapour or steam. A gas consists of a vast number of molecules which fly about entirely independently of one another, except at the rare instants at which two collide, and so interfere with each other's motion. The extent to which the molecules interfere with one another must obviously depend on their sizes; the larger they are, the more frequent their collisions will be, and the more they will interfere with one another's motion. Actually the extent of this interference provides the best means of estimating the sizes of molecules. They prove to be exceedingly small, being for the most part about a hundred-millionth of an inch in diameter, and, as a general rule, the simpler molecules have the smaller diameters, as we should expect. The molecule of water has a diameter of 1.8 hundred-millionths of an inch (4.6×10^{-8} cms.), while that of the simpler hydrogen molecule is only just over a hundred-millionth of an inch (2.7×10^{-8} cms.). The fact that a number of different lines of investigation all attribute the same diameters to these molecules provides an excellent proof of the reality of their existence.

As molecules are so exceedingly small, they must also be exceedingly numerous. A pint of water contains 1.89×10^{25} molecules, each weighing 1.06×10^{-24} ounces. If these molecules were placed end to end, they would form a chain capable of encircling the earth over 200 million times. If they were scattered over the whole land surface of the earth, there would be nearly 100 million molecules to every square inch of land. If we think of the molecules as tiny seeds, the total amount of seed needed to sow the whole earth at the rate of 100 million molecules to the square inch could be put into a pint pot.

These molecules move with very high speeds; in the ordinary air of an ordinary room, the average molecular speed is about 500 yards a second. This is roughly the speed of a rifle-bullet, and is rather more than the ordinary speed of sound. As we are familiar with this latter speed from everyday experience, it is easy to form some conception of molecular speeds in a gas. It is not a mere accident that molecular speeds are comparable with the speed of sound. Sound is a disturbance which one molecule passes on to another when it collides with it, rather like relays of messengers passing a message on to one another, or Greek torch-bearers handing on their lights. Between collisions the message is carried forward at exactly the speed at which the molecules travel. If these all travelled with precisely the same speed and in precisely the same direction, the sound would of course travel with just the speed of the molecules. But many of them travel on oblique courses, so that although the average speed of individual molecules in ordinary air is about 500 yards a second, the net forward velocity of the sound is only about 370 yards a second.

EXPLORING THE ATOM

At high temperatures the molecules may have even greater speeds; the molecules of steam in a boiler may move at 1000 yards a second.

It is the high speed of molecular motion that is responsible for the great pressure exerted by a gas; any surface in contact with the gas is exposed to a hail of molecules each moving with the speed of a rifle-bullet. For instance, the piston in a locomotive cylinder is bombarded by about 14×10^{28} molecules every second. This incessant fusillade of innumerable tiny bullets urges the piston forward in the cylinder, and so propels the train. With each breath we take, swarms of millions of millions of millions of molecules enter our bodies, each moving at about 500 yards a second, and nothing but their incessant hammering on the walls of our lungs keeps our chests from collapsing.

Perhaps the best general mental picture we can form of a gas is that of an incessant hail of shot or rifle-bullets flying indiscriminately in all directions, and running into one another at frequent intervals. In ordinary air each molecule collides with some other molecule about 3000 million times every second, and travels an average distance of about 1/160,000 inch between successive collisions. If we compress a gas to a greater density, more molecules are crowded into a given space, so that collisions become more frequent and the molecules travel shorter distances between collisions. If, on the contrary, we reduce the pressure of the gas, and so lessen its density, collisions become less frequent and the distance of travel of a molecule between successive collisions —the "free-path" as it is called—is increased. In the lowest vacua which are at present obtainable in the laboratory, a molecule can travel over 100 yards without colliding with any other molecules, although there are still 600,000 million molecules to the cubic inch.

Under astronomical conditions still lower vacua may occur. In some nebulae molecules of gas may travel millions of miles without a collision, so few are the molecules to a given volume of space.

It might be thought that the flying molecules would soon be brought to rest by their collisions; rifle-bullets undoubtedly would, but not the molecule bullets of a gas, for reasons now to be explained.

Energy. The amount of the charge of powder used to fire a rifle-bullet gives a measure of the "energy of motion" which is imparted to the bullet. To fire a bullet of double weight requires twice as much powder, because the energy of motion of a bullet, or indeed of any other moving body, is proportional to its weight. But to fire the same bullet with double speed does not merely require double the charge of powder. Four times as much powder is needed, because the energy of motion of a moving body is proportional to the *square* of its speed. The experienced motorist is familiar with this; if our brakes stop our car in 10 feet when we are travelling 10 miles an hour, they will not stop it in 20 feet when travelling at 20 miles an hour; we need 40 feet. Double speed requires four times the distance to pull up in, because double speed represents fourfold energy of motion. In general, the energy of motion of any moving body whatever is proportional both to the weight of the body and to the square of its speed.*

* This is expressed in the mathematical formula $\frac{1}{2}mv^2$ for the energy of motion of a body of weight m moving with a speed v. If m is measured in grammes, and v in centimetres per second, the energy of motion of the body is said to be $\frac{1}{2}mv^2$ "ergs." Thus an "erg" is the energy of motion of a body of 2 grammes weight (so that $\frac{1}{2}m=1$) moving with a speed of one centimetre a second. As an example, the energy of an express train of 300 tons' weight (3×10^8 gms.) moving at 60 miles an hour (2682 cms. a second) is 1079×10^{14} ergs; a cannon-ball or shell weighing a ton and moving at 1520 feet a second has precisely the same energy.

One of the great achievements of nineteenth-century physics was to establish the general principle known as the "conservation of energy." Energy can exist in a number of forms, and can change about almost endlessly from one form to another, but it can never be utterly destroyed The energy of a moving body is not lost when the body is brought to rest, it merely takes some other form. When a bullet is brought to rest by hitting a target, part of its energy of motion goes into heating up the target, and part into heating up, or perhaps even melting, the bullet. In its new guise of heat, there is just as much energy as there was in the original motion of the bullet.

In accordance with the same principle, energy cannot be created; all existing energy must have existed from all time, although possibly in some form entirely different from its present form. For instance, gunpowder contains a large amount of energy stored up in the form of chemical energy; we have to take precautions to prevent this bottled-up energy suddenly breaking free and doing damage, as, for instance, by exploding the vessel in which it is contained, kicking things up into the air, and so forth. A rifle is in effect a device for setting free the energy contained in a measured charge of gunpowder, and directing as much as possible of it into the form of energy of motion of a bullet. When we fire a bullet at a target, a specified amount of energy (determined by the charge of powder we have used) is transformed from chemical energy, residing in the powder, first into energy of motion, residing in the bullet (and to a minor degree in the recoil of the rifle), and then finally into heat energy, residing partly in the spent bullet and partly in the target. Here we have energy taking three different forms in rapid succession. All the life of the universe may be regarded as manifestations of energy masquerading in various forms, and all the changes in the universe as energy running about from one of these forms to the other, but always without altering its total amount. Such is the great law of conservation of energy.

Among the commoner forms of energy may be mentioned electric energy, as exemplified by the energy of a charged accumulator or of a thundercloud: mechanical energy, as exemplified in the coiled spring of a watch or the raised weight of a clock: chemical energy, as exemplified by the energy stored up in gunpowder or in coal, wood and oil: energy of motion, as exemplified by the motion of a bullet, and finally heat energy, which, as we have seen, is exemplified by the heat which appears when the motion of a rifle-bullet is checked.

Heat. Let us examine further into heat as a possible form of energy. When we want to warm a room, we light a fire and set free some of the chemical energy which is stored up in coal or wood, or we turn on an electric heater and let the electric current transport to us some of the energy which is being set free by the burning of coal in a distant power-station. But what, in the last resort, is heat, and how does it come to be a mode of energy?

Heat, whether of a gas, a liquid or a solid, is merely the energy of motion of individual molecules. When we heat up the air of a room we simply make its molecules move faster, and the total heat of the substance is the total energy of all the molecules of which it is composed. In pumping up a bicycle tyre, we drive the piston of the pump forward in opposition to the impact of innumerable millions of molecules of air inside the pump. In kicking the opposing molecules out of its way, the piston increases their speed of motion. The resulting increase in the energy of motion of the molecules is simply an increase of heat. We could verify this by inserting a thermometer, or, still more simply, by putting our hand on the pump; it feels hot.

The molecules of a solid are not possessed of much energy, and so do not move very fast—so slowly indeed that they seldom change their relative positions, the neighbouring molecules gripping them so firmly that their feeble energy of motion cannot extricate them. If we warm the solid up, the molecules acquire more energy, and so begin to move faster. After a time they are moving with such speeds that they can laugh at the restraining pulls from their neighbours; each molecule has enough energy of motion to go where it pleases, and we have a crowd of molecules moving freely as independent units, jostling one another and pushing their way past one another; the substance has assumed the liquid state. To make the picture definite, ice has melted and become water; the frozen grip is relaxed, and the molecules flow freely past one another. Each still exerts forces on its neighbours, but these are no longer strong enough to preclude all motion. Heat the liquid further, thus further increasing the energy of motion of the molecules, and these begin to break loose entirely from their bonds and fly about freely in space forming a gas or vapour. If we go on supplying heat, the whole substance will in time assume the gaseous state. Heating the gas still further merely causes the molecule-bullets to fly faster; it increases their energy of motion.

The average energy of motion of the molecules in a gas is proportional to the temperature of the gas—indeed, this is the way in which temperature is defined. The temperature must not, however, be measured on the Fahrenheit or

Centigrade scale in ordinary use, but on what is called the "absolute" scale, which has its zero at $-273°$ Centigrade, or $-469°$ Fahrenheit. This "absolute" zero, being the temperature of a body which has no further heat to lose, is the lowest temperature possible. We can approach to within two or three degrees of it in the laboratory, and find that it freezes air, hydrogen and even helium, the most refractory gas of all, solid. A thermometer placed out in interstellar space, far from any star, would probably shew a temperature of only about four degrees above absolute zero, while still lower temperatures must be reached out beyond the limits of the galactic system.

Molecular Collisions. We may now try to picture a collision between two molecule-bullets in a gas. Lead bullets colliding on a battlefield would probably change most of their energy of motion into heat-energy; they would become hotter, or perchance even melt. But how can the molecule-bullets transform their energy of motion into heat-energy? For them heat and energy of motion are not two different forms of energy, they are one and the same thing; their heat is their energy of motion. The total energy must be conserved, and there is no new disguise that it can assume. So it comes about that when two molecule-bullets collide, the most that can happen is that they may exchange a certain amount of energy of motion. If their energies of motion before collision were, say, 7 and 5 respectively, their energies after collision may be 6 and 6, or 8 and 4, or 9 and 3, or any other combination which adds up to 12.

It is the same at every collision; energy can neither be lost nor transformed, and so the bullets on the molecular battlefield go on flying for ever, happily hitting only one another, and doing no harm to one another when they hit. Their energies of motion go up and down, down and up, according as they make lucky hits or the reverse, but the most they have to fear are fluctuations and never total loss of energy; their motion is perpetual.

Atoms

In the gaseous state, each separate molecule retains all the chemical properties of the solid or liquid substance from which it originated; molecules of steam, for instance, moisten salt or sugar, or combine with thirsty substances such as unslaked lime or potassium chloride, just as water does.

Is it possible to break up the molecules still further? Lucretius and his predecessors would, of course, have said: "No." A simple experiment, which, however, was quite beyond their range, will speedily shew that they were wrong.

On sliding the two wires of an ordinary electric bell circuit into a tumbler of water, down opposite sides, bubbles of gas will be found to collect on the wires, and chemical examination shews that the two lots of gas have entirely different properties. They cannot, then, both be water-vapour, and in point of fact neither of them is; one proves to be hydrogen and the other oxygen. There is found to be twice as much hydrogen as oxygen, whence we conclude that the electric current has broken up each molecule of water into two parts of hydrogen and one of oxygen. These smaller units into which a molecule is broken are called "atoms." Each molecule of water consists of two atoms of hydrogen (H) and one atom of oxygen (O); this is expressed in its chemical formula H_2O.

All the innumerable substances which occur on earth—shoes, ships, sealing-wax, cabbages, kings, carpenters, walruses, oysters, everything we can think of—can be analysed into their constituent atoms, either in this or in other ways. It might be thought that a quite incredible number of different kinds of atoms would emerge from the rich variety of substances we find on earth. Actually the number is quite small. The same atoms turn up again and again, and the great variety of substances we find on earth result, not from any great variety of atoms entering into their composition, but from the great variety of ways in which a few types of atoms can be combined—just as in a colour-print three colours can be combined so as to form almost all the colours we meet in nature, not to mention other weird hues such as never were on land or sea.

Analysis of all known terrestrial substances has, so far, revealed only 90 different kinds of atoms. Probably 92 exist, there being reasons for thinking that two, or possibly even more, still remain to be discovered. Even of the 90 already known, the majority are exceedingly rare, most common substances being formed out of the combinations of about 14 different atoms, say hydrogen (H), carbon (C), nitrogen (N), oxygen (O), sodium (Na), magnesium (Mg), aluminum (Al), silicon (Si), phosphorus (P), sulphur (S), chlorine (Cl), potassium (K), calcium (Ca), and iron (Fe).

In this way, the whole earth, with its endless diversity of substances, is found to be a building built of standard bricks —the atoms. And of these only a few types, about 14, occur at all abundantly in the structure, the others appearing but rarely.

Spectroscopy. Just as a bell struck with a hammer emits a characteristic note, so every atom put in a flame or in an electric arc or discharge-tube, emits a characteristic light. When Newton passed sunlight through a prism, he found it to be a blend of all the colours of the rainbow. In the same way the modern spectroscopist, with infinitely more refined instruments, can analyse any light into all the constituent colours which enter into its composition. The rainbow of colours so produced—the "spectrum"—is crossed by the pattern of light or dark lines or bands which the astronomer utilises to determine the speeds of recession or approach of the stars. From an examination of this pattern the skilled spectroscopist can at once announce the type of atom from which the light emanates, so much so that one of the most delicate tests for the presence of certain substances is the spectroscopic test.

This spectroscopic method of analysis is by no means confined to terrestrial substances. In 1814 Fraunhofer repeated Newton's analysis of sunlight, and found its spectrum to be crossed by certain dark lines, still known as Fraunhofer lines. The spectroscopist has no difficulty in interpreting these dark lines; they indicate the presence in the sun of the common terrestrial elements, hydrogen, sodium, calcium, and iron. For reasons which we shall see later (p. 118 below), the atoms of these substances drink up the light of precisely those colours which the Fraunhofer lines shew to be absent from the solar spectrum. This spectrum is now known to be incomparably more intricate than Fraunhofer thought, but practically all the lines which occur in it can be assigned to atoms known on earth, and the same is true of the spectra of all the stars in the sky. It is tempting to jump to the generalisation that the whole universe is built solely of the 90 or 92 types of atoms found on earth, but at present there is no justification for this. The light we receive from the sun and stars comes only from the outermost layers of their surfaces, and so conveys no information at all as to the types of atoms to be found in the stars' interiors. Indeed we have no knowledge of the types of atoms which occur in the interior of our own earth.

The Structure of the Atom. Until quite recently, atoms were regarded as the permanent bricks of which the whole universe was built. All the changes of the universe were supposed to amount to nothing more drastic than a re-arrangement of permanent indestructible atoms; like a child's box of bricks, these built many buildings in turn. The story of twentieth-century physics is primarily the story of the shattering of this concept.

It was towards the end of the last century that Crookes, Lenard, and, above all, Sir J. J. Thomson first began to break up the atom. The structures which had been deemed the unbreakable bricks of the universe for more than 2000 years, were suddenly shown to be very susceptible to having fragments chipped off. A mile-stone was reached in 1895, when Thomson shewed that these fragments were identical, no matter what type of atom they came from; they were of equal weight and they carried equal charges of negative electricity. On account of this last property they were called "electrons." The atom cannot, however, be built up of electrons and nothing else, for as each electron carries a negative charge of electricity, a structure which consisted of nothing but electrons would also carry a negative charge. Two negative charges of electricity repel one another, as also do two positive charges, while two charges, one of positive and one of negative electricity, attract one another. This makes it easy to determine whether any body or structure carries a positive or a negative charge of electricity, or no charge at all. Observation shews that a complete atom carries no charge at all, so that somewhere in the atom there must be a positive charge of electricity, of amount just sufficient to neutralise the combined negative charges of all the electrons.

In 1911 experiments by Sir Ernest Rutherford and others revealed the architecture of the atom. As we shall soon see (p. 105 below), nature herself provides an endless supply of small particles charged with positive electricity, and moving with very high speeds, in the α-particles shot off from radio-active substances. Rutherford's method was in brief to fire these into atoms and observe the result. And the surprising result he obtained was that the vast majority of these bullets passed straight through the atom as though it simply did not exist. It was like shooting at a ghost.

Yet the atom was not all ghostly. A tiny fraction—perhaps one in 10,000—of the bullets were deflected from their courses as if they had met something very substantial indeed. A mathematical calculation shewed that these obstacles could only be the missing positive charges of the atoms.

A detailed study of the paths of these projectiles proved that the whole positive charge of an atom must be concentrated in a single very small space, having dimensions of the order of only a millionth of a millionth of an inch. In this way, Rutherford was led to propound the view of atomic structure which is generally associated with his name. He supposed the chemical properties and nature of the atom to reside in a weighty, but excessively minute, central "nucleus" carrying a positive charge of electricity, around which a number of negatively charged electrons described orbits. It was of course necessary to suppose the electrons to be in motion in the atom, otherwise the attraction of positive for negative electricity would immediately draw them into the central nucleus—just as gravitational attraction would cause the earth to fall into the sun, where it not for the orbital motion of the former. In brief Rutherford supposed the atom to be constructed like the solar system, the heavy central nucleus playing the part of the sun and the electrons acting the parts of the planets.

The speeds with which these electrons fly round their tiny orbits are terrific. The average electron revolves around its nucleus several thousand million million times every second, with a speed of hundreds of miles a second. Thus the smallness of their orbits does not prevent the electrons moving with higher orbital speeds than the planets, or even the stars themselves.

By clearing a space around the central nucleus, and so preventing other atoms from coming too near to it, these electronic orbits give size to the atom. The volume of space kept clear by the electrons is enormously greater than the total volume of the electrons; roughly, the ratio of volumes is that of the battlefield to the bullets. The atom, with a radius of about 2×10^{-8} cms., has about 100,000 times the diameter, and so about a thousand million million times the volume, of a single electron, which has a radius of only about 2×10^{-13} cms. The 'nucleus,' although it generally weighs 3000 or 4000 times as much as all the electrons in the atom together, is at most comparable in size with, and may be even smaller than, a single electron.

We have already commented on the extreme emptiness of astronomical space. Choose a point in space at random, and the odds against its being occupied by a star are enormous. Even the solar system consists overwhelmingly of empty space; choose a spot inside the solar system at random, and there are still immense odds against its being occupied by a planet or even by a comet, meteorite or smaller body. And now we see that this emptiness extends also to the space of physics. Even inside the atom we choose a point at random, and the odds against there being anything there are immense; they are of the order of at least millions of millions to one. We saw how six specks of dust inside Waterloo Station represented—or rather over-represented—the extent to which space was crowded with stars. In the same way a few wasps—six for the atom of carbon—flying around in Waterloo Station will represent the extent to which the atom is crowded with electrons—all the rest is emptiness. As we pass the whole structure of the universe under review, from the giant nebulae and the vast interstellar and internebular spaces down to the tiny structure of the atom, little but vacant space passes before our mental gaze. We live in a gossamer universe; pattern, plan and design are there in abundance, but solid substance is rare.

Atomic Numbers. The number of electrons which fly round in orbits in an atom is called the "atomic number" of the atom. Atoms of all atomic numbers from 1 to 92 have been found, except for two missing numbers 85 and 87. As already mentioned, it is highly probable that these also exist, and that there are 92 "elements" whose atomic numbers occupy the whole range of atomic numbers from 1 to 92 continuously.

The atom of atomic number unity is of course the simplest of all. It is the hydrogen atom, in which a solitary electron revolves around a nucleus whose charge of positive electricity is exactly equal in amount, although opposite in sign, to the charge on the negative electron.

Next comes the helium atom of atomic number 2, in which two electrons revolve about a nucleus which has four times the weight of the hydrogen nucleus, although carrying only twice its electric charge. After this come the lithium atom of atomic number 3, in which three electrons revolve around a nucleus having six times the weight of the hydrogen atom and three times its charge. And so it goes on, until we reach uranium, the heaviest of all atoms known on earth, which has 92 electrons describing orbits about a nucleus of 238 times the weight of the hydrogen nucleus.

Radio-activity

While it was still engaged in breaking up the atom into its component factors, physical science was beginning to discover that the nuclei themselves were neither permanent nor indestructible. In 1896, Becquerel had found that various substances containing uranium possessed the remarkable property, as it then appeared, of spontaneously affecting photographic plates in their vicinity. This observation led to the discovery of a new property of matter, namely radio-activity. All the results obtained from the study of radio-activity in the few following years were co-ordinated in the hypothesis of "spontaneous disintegration" which Rutherford and Soddy advanced in 1903. According to this hypothesis in its present form, radio-activity indicates a spontaneous break-up of the nuclei of the atoms of radio-active substances. These atoms are so far from being permanent and indestructible that their very nuclei crumble away with the mere lapse of time, so that what was once the nucleus of a uranium atom is transformed, after sufficient time, into the nucleus of a lead atom.

The process of transformation is not instantaneous; it proceeds gradually and by distinct stages. During its progress, three types of product are emitted, which are designated α-rays, β-rays, and γ-rays.

These were originally described as rays because they have

the power of penetrating through a certain thickness of air, metal, or other substance. Their true nature was discovered later. It is well known that magnetic forces such as, for instance, occur in the space between the poles of a magnet, cause a moving particle charged with electricity to deviate from a straight course; it deviates in one direction or the other according as it is charged with positive or negative electricity. On passing the various rays emitted by radio-active substances through the space between the poles of a powerful magnet, the α-rays were found to consist of particles charged with positive electricity, and the β-rays to consist of particles charged with negative electricity. But the most powerful magnetic forces which could be employed failed to cause the slightest deviation in the paths of the γ-rays, from which it was concluded that either the γ-rays were not material particles at all, or that, if they were, they carried no electric charges. The former of these alternatives was subsequently proved to be the true one.

α *particles.* The positively charged particles which constitute α-rays are generally described as α-particles. In 1909 Rutherford and Royds allowed α-particles to penetrate through a thin glass wall of less than a hundredth of a millimetre thickness into a chamber from which they could not escape—a sort of mouse-trap for α-particles. They found that so long as the number of α-particles in the vessel went on increasing, an accumulation of helium was forming. In this way it was established that the positively charged α-particles are simply nuclei of helium atoms.

These particles move with enormous speeds, which depend upon the nature of the radio-active substance from which they have been shot out. The fastest of all, those emitted by Thorium C', move with a speed of 12,800 miles a second;

even the slowest, those from Uranium 1, have a speed of 8800 miles a second, which is about 30,000 times the ordinary molecular velocity in air. Particles moving with these speeds knock all ordinary molecules out of their way; this explains the great penetrating power of the α-rays.

β-*particles*. By examining their motion under magnetic forces, the β-rays were found to consist of negatively charged electrons, exactly similar to those which revolve orbitally in all atoms. As an α-particle carries a positive charge equal in amount to that of two electrons, an atom which has ejected an α-particle is left with a deficiency of positive charge, or what comes to the same thing, with a negative charge, equal to that of two electrons. Consequently it is natural, and indeed almost inevitable, that the ejections of α-particles should alternate with an ejection of negatively charged electrons, so that the balance of positive and negative electricity in the atom may be maintained. The β-particles, move with even greater speeds than the α-particles, many approaching to within a few per cent. of the velocity of light (186,000 miles a second).

One of the most beautiful devices known to physical science, the invention of Professor C. T. R. Wilson, makes it possible to study the motions of the α- and β-particles as they thread their way through a gas, colliding with its molecules on their way. A chamber through which the particles are made to travel is filled with water-vapour in such a condition that the passage of an electrically charged particle leaves behind it a trail of condensations which can be photographed. As an example, Plate XIII shews a photograph taken by Professor Wilson himself, in which the trails of both α- and β-particles appear on the same plate. As the α-particles weigh about 7400 times as much as the β-particles,

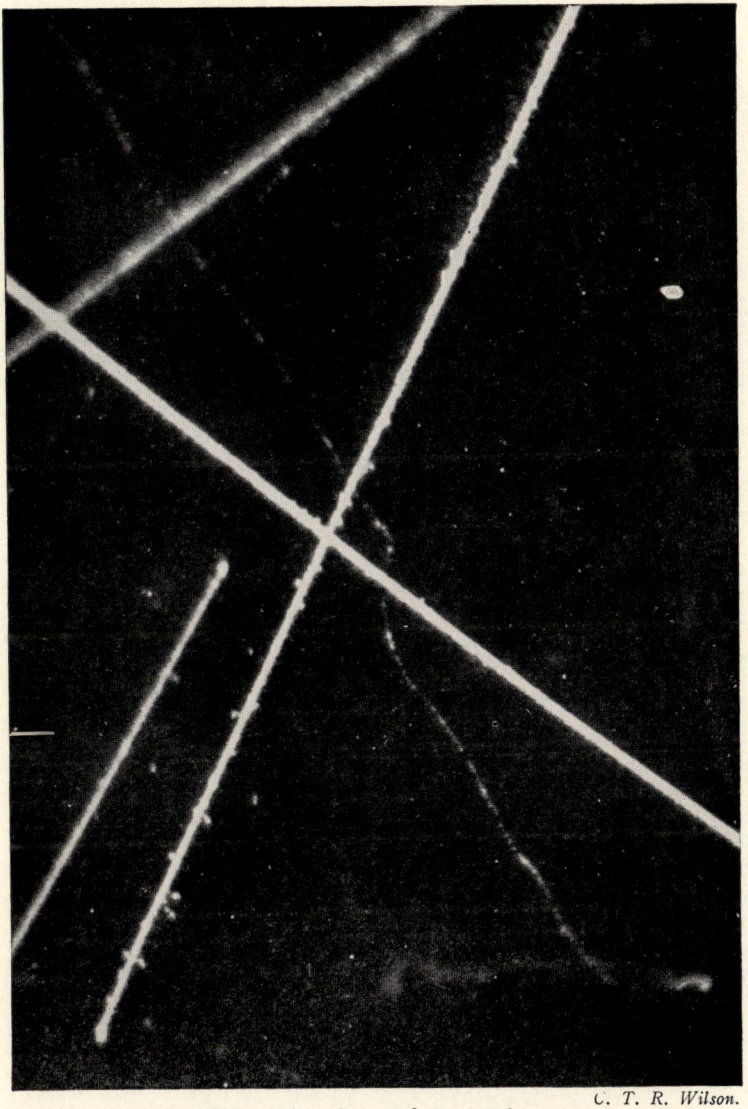

PLATE XIII

The tracks of α- and β-particles.

C. T. R. Wilson.

they naturally create more disturbance in the gas, and so leave broader and more pronounced tracks; also they pursue a comparatively straight course while the lighter β-particles are deflected from their courses by many of the molecules they meet. The plate shews four α-particle tracks and one (much fainter) β-ray track. The knobby-looking projections which may be seen on one of the α-ray tracks are of interest; they represent the short paths of electrons knocked out of atoms by the passage of the α-particle.*

γ-rays. The γ-rays are not material particles at all; they prove to be merely radiation of a very special kind, which we shall now discuss.

Radiation

Disturb the surface of a pond with a stick and a series of ripples starts from the stick and travels, in a series of ever-expanding circles, over the surface of the pond. As the water resists the motion of the stick, we have to work to keep the pond in a state of agitation. The energy of this work is transformed, in part at least, into the energy of the ripples. We can see that the ripples carry energy about with them, because they cause a floating cork or a toy boat to rise up against the earth's gravitational pull. Thus the ripples provide a mechanism for distributing over the surface of the pond the energy that we put into the pond through the medium of the moving stick.

Light and all other forms of radiation are analogous to water-ripples or waves, in that they distribute energy from a central source. The sun's radiation distributes through space the vast amount of energy which is generated inside the sun. We hardly know whether there is any actual wave-

* These were called δ-rays by Bumstead.

motion in light or not, but we know that light, as well as other types of radiation, are propagated in such a form that they have some of the properties of a succession of waves.

We have seen how the different colours of light which in combination constitute sunlight can be separated out by passing the light through a prism. An alternative instrument, the diffraction-grating, analyses light into its constituent wave-lengths,* and these are found to correspond to the different colours of the rainbow. This shews that different colours of light represent different wave-lengths, and at the same time provides a means of measuring the actual wave-lengths of light of different colours. These prove to be very minute. The reddest light we can see, which is that of longest wave-length, has a wave-length of only 3/100,000 inch (7.5×10^{-5} cms.); the most violet light we can see has a wave-length only half of this, or 0.000015 inch. Light of all colours travels with the same uniform speed of 186,000 miles, or 3×10^{10} centimetres, a second. The number of waves of red light which pass any fixed point in a second is accordingly no fewer than four hundred million million. This is called the "frequency" of the light. Violet light has the still higher frequency of eight hundred million million; when we see violet light, eight hundred million million waves of light enter our eyes each second.

The spectrum of analysed sunlight appears to the eye to stretch from red light at one end to violet light at the other, but these are not its true limits. If certain chemical salts are placed beyond the violet end of the visible spectrum, they are found to shine vividly, shewing that even out here energy is being transported, although in invisible form.

* The wave-length in a system of ripples is the distance from the crest of one ripple to that of the next, and the term may be applied to all phenomena of an undulatory nature.

EXPLORING THE ATOM

Regions of invisible radiation stretch indefinitely from both ends of the visible spectrum. From one end—the red—we can pass continuously to waves of the type used for wireless transmission, which have wave-lengths of the order of hundreds, or even thousands, of yards. From the violet end, we pass through waves of shorter and ever shorter wave-length—all the various forms of ultra-violet radiation. At wave-lengths of from about a hundredth to a thousandth of the wave-length of visible light, we come to the familiar X-rays, which penetrate through inches of our flesh, so that we can photograph the bones inside. Far out even beyond these, we come to the type of radiation which constitutes the γ-rays, its wave-length being of the order of 1/10,000,000,000 inch, or only about a hundred-thousandth part of the wave-length of visible light. Thus the γ-rays may be regarded as invisible radiation of extremely short wave-length. We shall discuss the exact function they serve later. For the moment let us merely remark that in the first instance they served the extremely useful function of fogging Becquerel's photographic plates, thus leading to the detection of the radio-active property of matter.

Thus we see that the break-up of a radio-active atom may be compared to the discharge of a gun; the α-particle is the shot fired, the β-particles are the smoke, and the γ-rays are the flash. The atom of lead which finally remains is the unloaded gun, and the original radio-active atom, of uranium or what not, was the loaded gun. And the special peculiarity of radio-active guns is that they go off spontaneously and of their own accord. All attempts to pull the trigger have so far failed, or at least have led to inconclusive results; we can only wait, and the gun will be found to fire itself in time.

Atomic Nuclei

With the unimportant exceptions of potassium and rubidium (of atomic numbers 19 and 37), the property of radio-activity occurs only in the most complex and massive of atoms, being indeed confined to those of atomic numbers above 83. Yet, although the lighter atoms are not liable to spontaneous disintegration in the same way as the heavy radio-active atoms, their nuclei are of composite structure, and can be broken up by artificial means. In 1920, Rutherford, using radio-active atoms as guns, fired α-particles at light atoms and found that direct hits broke up their nuclei. There is, however, found to be a significant difference between the spontaneous disintegration of the heavy radio-active atoms, and the artificial disintegration of the light atoms; in the former case, apart from the ever-present β-rays and γ-rays, only α-particles are ejected, while in the latter case α-particles were not ejected at all, but particles of only about a quarter of their weight, which proved to be identical with the nuclei of hydrogen atoms.

These sensational events in the atomic underworld can be photographed by Professor C. T. R. Wilson's condensation method already explained. Plate XIV shews two collisions of an α-particle with a nitrogen atom photographed by Mr. P. M. S. Blackett. The straight lines are merely the quite uneventful tracks of ordinary α-particles similar to those already shewn in Plate XIII. But one α-particle track in each photograph suddenly branches, so that the complete figure is of a Y-shape.

There is little room for doubt that in fig. 1 the branch occurs because the α-particle has collided with a nitrogen atom; the lower branch of the Y is the track of the α-particle before the collision; the two upper branches are the tracks

PLATE XIV

Fig. 1.

Fig. 2.

P. M. S. Blackett.

Collision of α-particles with Nitrogen Atoms.

In Fig. 1 the α-particle merely rebounds from a nitrogen atom. In Fig. 2 it drives out a proton and then joins itself to the atom.

of the α-particle and the nitrogen atom after the collision, the latter now moving with enormous speed and hitting everything out of its way. By taking simultaneous photographs in two directions at right angles, as shewn in the Plate, Mr. Blackett was able to reconstruct the whole collision, and the angles were found to agree exactly with those which dynamical theory would require on this interpretation of the photograph.

The occurrence photographed in fig. 2 is of a different type from that seen in fig. 1, for the angles do not agree with those which dynamical theory would require if the upper branches of the Y were the tracks of the α-particles and the nitrogen atom as in fig. 1. The lower branch of the Y is still an ordinary α-particle track, but the upper branch on the left is the track of a particle smaller than an α-particle, namely a particle of quarter-weight shot out of the nucleus, whilst the fork to the left is that of the nitrogen atom moving along in company with the α-particle, which it has captured. It would take too much space to describe in full the beautiful method by which Blackett has established this interpretation of his photographs, but there is little room for doubt that in fig. 2 he has actually succeeded in photographing the break-up of the nucleus of an atom of nitrogen.

Isotopes. Two atoms have the same chemical properties if the charges of positive electricity carried by their nuclei are the same. The amount of this charge fixes the number of electrons which can revolve around the nucleus, this number being of course exactly that needed to neutralise the electric field of the nucleus, and this in turn fixes the atomic number of the element. But Dr. Aston has shewn that atoms of the same chemical element, say neon or chlorine, may have nuclei of different weights. The various forms which the atoms of the same chemical element can assume are known as isotopes, being of course distinguished by their different weights. Aston further made the highly significant discovery that the weights of all atoms are, to a very close approximation, multiples of a single definite weight. This unit weight is approximately equal to the weight of the hydrogen atom, but is more nearly equal to a sixteenth of the weight of the oxygen atom. The weight of any type of atom, measured in terms of this unit, is called the "atomic weight" of the atom.

Protons and Electrons. In conjunction with the results of Rutherford's artificial disintegration of atomic nuclei, Aston's results have led to the general acceptance of the hypothesis that the whole universe is built up of only two kinds of ultimate bricks, namely, electrons and protons. Each proton carries a positive charge of electricity exactly equal in amount to the negative charge carried by an electron, but has about 1840 times the weight of the electron. Protons are supposed to be identical with the nucleus of the hydrogen atom, all other nuclei being composite structures in which both protons and electrons are closely packed together. For instance, the nucleus of the helium atom, or α-particle, consists of four protons and two electrons, these giving it approximately four times the weight of the hydrogen atom, and a resultant charge equal to twice that of the nucleus of the hydrogen atom.

Yet this is not quite the whole story. If it were, every complete atom would consist of a certain number N of protons, together with just enough electrons, namely N, to neutralise the electric charges on the N protons, so that its ingredients would be precisely the same as those of N hydrogen atoms. Thus the weight of every atom would be an exact multiple of the weight of a hydrogen atom. Experiment shews this not to be the case.

Electromagnetic Energy. To get at the whole truth, we have to recognise that, in addition to containing material electrons and protons, the atom contains yet a third ingredient which we may describe as electromagnetic energy. We may think of this, although with something short of absolute scientific accuracy, as bottled radiation.

It is a commonplace of modern electromagnetic theory that radiation of every kind carries weight about with it, weight which is in every sense as real as the weight of a ton of coal. A ray of light causes an impact on any surface on which it falls, just as a jet or water does, or a blast of wind, or the fall of a ton of coal; with a sufficiently strong light one could knock a man down just as surely as with the jet of water from a fire-hose. This is not a mere theoretical prediction. The pressure of light on a surface has been both detected and measured by direct experiment. The experiments are extraordinarily difficult because, judged by all ordinary standards, the weight carried by radiation is exceedingly small; all the radiation emitted from a 50 horse-power searchlight working continuously for a century weighs only about a twentieth of an ounce.

It follows that any substance which is emitting radiation must at the same time be losing weight. In particular, the disintegration of any radio-active substance must involve a decrease of weight, since it is accompanied by the emission of radiation in the form of γ-rays. The ultimate fate of an ounce of uranium may be expressed by the equation:

$$1 \text{ ounce uranium} = \begin{cases} 0.8653 \text{ ounce lead,} \\ 0.1345 \text{ '' helium,} \\ 0.0002 \text{ '' radiation.} \end{cases}$$

The lead and helium together contain just as many electrons and just as many protons as did the original ounce of uranium, but their combined weight is short of the weight of the original uranium by about one part in 4000. Where 4000 ounces of matter originally existed, only 3999 now remain; the missing ounce has gone off in the form of radiation.

This makes it clear that we must not expect the weights of the various atoms to be exact multiples of the weight of the hydrogen atom; any such expectation would ignore the weight of the bottled-up electromagnetic energy which is capable of being set free and going off into space in the form of radiation as the atom changes its make up. The weight of this energy is relatively small, so that the weights of the atoms may be expected to be approximately integral multiples of that of the hydrogen atom, and this expectation is confirmed, but they will not be so exactly. The exact weight of our atomic building is not simply the total weight of all its bricks; something must be added for the weight of the mortar—the electromagnetic energy—which keeps the bricks bound together.

Thus the normal atom consists of protons, electrons, and energy, each of which contributes something to its weight. When the atom re-arranges itself, either spontaneously or under bombardment, protons and electrons may be shot off in the form of material particles (α- and β-rays) and energy may also be set free in the form of radiation. This radiation may either take the form of γ-rays, or, as we shall shortly see, of other forms of visible and invisible radiation. The final weight of the atom will be obtained by deducting from its original weight not only the weight of all the ejected electrons and protons, but also the weight of all the energy which has been set free as radiation.

QUANTUM THEORY

The series of concepts which we now approach are difficult to grasp and still more difficult to explain, largely, no doubt, because our minds receive no assistance from our everyday experience of nature.* It becomes necessary to speak mainly in terms of analogies, parables and models which can make no claim to represent ultimate reality; indeed it is rash to hazard a guess even as to the direction in which ultimate reality lies.

The laws of electricity which were in vogue up to about the end of the nineteenth century—the famous laws of Maxwell and Faraday—required that the energy of an atom should continually decrease, through the atom scattering energy abroad in the form of radiation, and so having less and less left for itself. These same laws predicted that all energy set free in space should rapidly transform itself into radiation of almost infinitesimal wave-length. Yet these things simply did not happen, making it obvious that the then prevailing electrodynamical laws had to be given up.

Cavity-radiation. A crucial case of failure was provided by what is known as "cavity-radiation." A body with a cavity in its interior is heated up to incandescence; no notice is taken of the light and heat emitted by its outer surface, but the light imprisoned in the internal cavity is let out through a small window and analysed into its constituent colours by a spectroscope or diffraction grating. It is this

---

* The reader whose interest is limited to astronomy may prefer to proceed at once to Chapter III.

radiation that is known as "cavity-radiation." It represents the most complete form of radiation possible, radiation from which no colour is missing, and in which every colour figures at its full strength. No known substance ever emits quite such complete radiation from its surface, although many approximate to doing so. We speak of such bodies as "full radiators."

The nineteenth-century laws of electromagnetism predicted that the whole of the radiation emitted by a full radiator or from a cavity ought to be found at or beyond the extreme violet end of the spectrum, independently of the precise temperature to which the body had been heated. In actual fact the radiation is usually found piled up at exactly the opposite end of the spectrum, and in no case does it ever conform to the predictions of the nineteenth century laws, or even begin to think of doing so.

In the year 1900, Professor Planck of Berlin discovered experimentally the law by which "cavity-radiation" is distributed among the different colours of the spectrum. He further shewed how his newly-discovered law could be deduced theoretically from a system of electromagnetic laws which differed very sensationally from those then in vogue.

Planck imagined all kinds of radiation to be emitted by systems of vibrators which emitted light when excited, much as tuning forks emit sound when they are struck. The old electrodynamical laws predicted that each vibration should gradually come to rest and then stop, as the vibrations of a tuning fork do, until the vibrator was in some way excited again. Rejecting all this, Planck supposed that a vibrator could change its energy by sudden jerks, and in no other way; it might have one, two, three, four or any other integral number of units of energy, but no intermediate fractional numbers, so that gradual changes of energy were rendered impossible. The vibrator, so to speak, kept no small change, and could only pay out its energy a shilling at a time until it had none left. Not only so, but it refused to receive small change, although it was prepared to accept complete shillings. This concept, sensational, revolutionary and even ridiculous, as many thought it at the time, was found to lead exactly to the distribution of colours actually observed in cavity-radiation.

In 1917, Einstein put the concept into the more precise form which now prevails. According to a theory previously advanced by Professor Niels Bohr of Copenhagen, an atomic or molecular structure does not change its configuration, or dissipate away its energy, by gradual stages. Gradualness is driven out of physics, and discontinuity takes its place. An atomic structure has a number of possible states or configurations which are entirely distinct and detached one from another, just as a weight placed on a staircase has only a possible number of positions; it may be 3 stairs up, or 4 or 5, but cannot be $3\frac{1}{4}$ or $3\frac{3}{4}$ stairs up. The change from one position to another is generally effected through the medium of radiation. The system can be pushed upstairs by absorbing energy from radiation which falls on it, or it may move downstairs to a state of lower energy and emit energy in the form of radiation in so doing. Only radiation of a certain definite colour, and so of a certain precise wavelength, is of any account for effecting a particular change of state. The problem of shifting an atomic system is like that of extracting a box of matches from a penny-in-the-slot machine; it can only be done by a special implement, to wit a penny, which must be of precisely the right size and weight —a coin which is either too small *or too large,* too light *or too heavy,* is doomed to fail. If we pour radiation of the wrong wave-length on to an atom, we may reproduce the comedy of the millionaire whose total wealth will not procure him a box of matches because he has not a loose penny, or we may reproduce the tragedy of the child who cannot obtain a slab of chocolate because its hoarded wealth consists of farthings and half-pence, but we shall not disturb the atom. When mixed radiation is poured on to a collection of atoms, these absorb the radiation of just those wave-lengths which are needed to change their internal states, and none other; radiation of all other wave-lengths passes by unaffected.

This selective action of the atom on radiation is put in evidence in a variety of ways; it is perhaps most simply shewn in the spectra of the sun and stars. Dark lines similar to those which Fraunhofer observed in the solar spectrum are observed in the spectra of practically all stars (see Plate VIII, p. 48), and we can now understand why this must be. Light of every possible wave-length streams out from the hot interior of a star, and bombards the atoms which form its atmosphere. Each atom drinks up that radiation which is of precisely the right wave-length for it, but has no interaction of any kind with the rest, so that the radiation which is finally emitted from the star is deficient in just the particular wave-lengths which suit the atoms. Thus the star shews an *absorption spectrum* of fine lines. The positions of these lines in the spectrum shew what types of radiation the stellar atoms have swallowed, and so enable us to identify the atoms from our laboratory knowledge of the tastes of different kinds of atoms for radiation. But what ultimately decides which types of radiation an atom will swallow, and which it will reject?

Planck had already supposed that radiation of each wave-length has associated with it a certain amount of energy, called the "quantum," which depends on the wave-length and on nothing else. The quantum is supposed to be proportional to the "frequency" (p. 108), or number of vibrations of the radiation per second,* and so is *inversely* proportional to the wave-length of the radiation—the shorter the wave-length, the greater the energy of the quantum, and conversely. Red light has feeble quanta, violet light has energetic quanta, and so on.

Einstein now supposes that radiation of a given type can effect an atomic or molecular change, only if the energy needed for the change is precisely equal to that of a single quantum of the radiation. This is commonly known as Einstein's law; it determines the precise type of radiation needed to work any atomic or molecular penny-in-the-slot mechanism.*

We notice that work which demands one powerful quantum cannot be performed by two, or indeed by any number whatever, of feeble quanta. A small amount of violet (high-frequency) light can accomplish what no amount of red (low-frequency) light can effect—a circumstance with which every photographer in painfully familiar; we can admit as much red light as we please without any damage being done, but even the tiniest gleam of violet light spoils our plates.

The law prohibits the killing of two birds with one stone,

* To be precise, if v is the frequency of the radiation, its quantum of energy is hv, where h is a universal constant of nature, known as Planck's constant. This constant is of the physical nature of energy multiplied by time; its numerical value is:
$$6.55 \times 10^{-27} \text{ ergs} \times \text{seconds.}$$
* In the form of an equation:
$$E_1 - E_2 = hv,$$
where E_1, E_2 are the energies of the material system before and after the change, v is the frequency of the radiation, and h is Planck's constant already specified.

as well as the killing of one bird with two stones; the whole quantum is used up in effecting the change, so that no energy from this particular quantum is left over to contribute to any further change. This aspect of the matter is illustrated by Einstein's photochemical law: "in any chemical reaction which is produced by the incidence of light, the number of molecules which are affected is equal to the number of quanta of light which are absorbed." Those who manage penny-in-the-slot machines are familiar with a similar law: "the number of articles sold is exactly equal to the number of coins in the machine."

If we think of energy in terms of its capacity for doing damage, we see that radiation of short wave-length can work more destruction in atomic structures than radiation of long wave-length. Radiation of sufficiently short wave-length may not only rearrange molecules or atoms; it may break up any atom on which it happens to fall, by shooting out one of its electrons, giving rise to what is known as photoelectric action. Again there is a definite limit of frequency, such that light whose frequency is below this limit does not produce any effect at all, no matter how intense it may be; whereas as soon as we pass to frequencies above this limit, light of even the feeblest intensity starts photoelectric action at once. Again the absorption of one quantum breaks up only one atom, and further ejects only one electron from the atom. If the radiation has a frequency above this limit, so that its quantum has more energy than the minimum necessary to remove a single electron from the atom, the whole quantum is still absorbed, the excess energy now being used in endowing the ejected electron with motion.

Electron Orbits. These concepts are based upon Bohr's supposition that only a limited number of orbits are open to the electrons in an atom, all others being prohibited for reasons which we still do not fully understand, and that an electron is free to move from one permitted orbit to another under the stimulus of radiation. Bohr himself investigated the way in which the various permitted orbits are arranged. Modern investigations indicate the need for a good deal of revision of his simple concepts, but we shall discuss these in some detail, partly because Bohr's picture of the atom still provides the best working mechanical model we have, and partly because an understanding of his simple theory is absolutely essential to the understanding of the far more intricate theories which are beginning to replace it.

The hydrogen atom, as we have already seen, consists of a single proton as central nucleus, with a single electron revolving around it. The nucleus, with about 1840 times the weight of the electron, stands practically at rest unagitated by the motion of the latter, just as the sun remains practically undisturbed by the motion of the earth round it. The nucleus and electron carry charges of positive and negative electricity, and therefore attract one another; this is why the electron describes an orbit instead of flying off in a straight line, again like the earth and sun. Furthermore, the attraction between electric charges of opposite sign, positive and negative, follows, as it happens, precisely the same law as gravitation, the attraction falling off as the inverse square of the distance between the two charges. Thus the nucleus-electron system is similar in all respects to a sun-planet system, and the orbits which an electron can describe around a central nucleus are precisely identical with those which a planet can describe about a central sun; they consist of a system of ellipses each having the nucleus in one focus (p. 44).

Yet the general concepts of quantum-dynamics prohibit the electron from moving in all these orbits indiscriminately. According to Bohr, the electron of the hydrogen atom can move in a certain number of circular orbits whose diameters are proportional to the squares of the natural numbers 1, 4, 9, 16, 25, . . .; it can also move in a series of elliptic orbits whose greatest diameters are respectively equal to the diameters of the possible circular orbits, although these elliptic

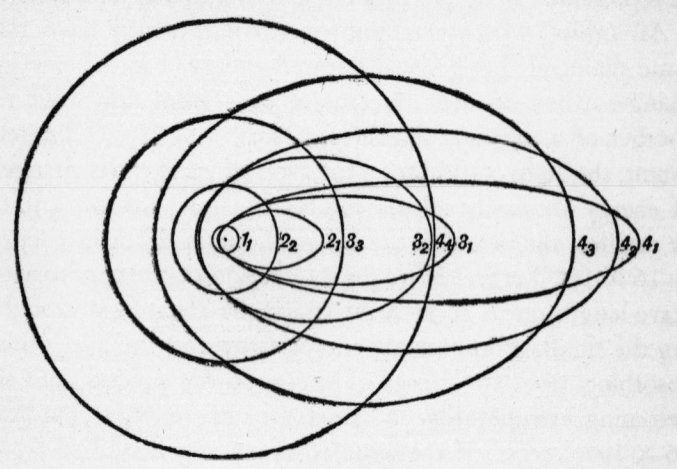

FIG. 10.—The arrangement of electron-orbits in the hydrogen atom (Bohr's model).

orbits are still further limited by the condition that their eccentricities must have certain definite values. All other orbits are in some way prohibited.

The smallest orbits which the electron can describe in the hydrogen atom are shewn in fig. 10. The smallest orbit of all, of diameter 1, is marked 1_1; beyond this come two orbits of diameter 4 marked 2_1, 2_2; then three orbits of diameter 9 marked 3_1, 3_2, 3_3; and four orbits of diameter 16 marked

4_1, 4_2, 4_3, 4_4. The diagram stops here for want of space, but the available orbits go on indefinitely. Even under laboratory conditions, electrons may move in orbits of a hundred times the diameter of that marked 1_1. Under the more rarified conditions of stellar atmospheres the hydrogen atom may swell out to even greater dimensions, and stellar spectra provide evidence of orbits having over a thousand times the dimensions of the 1_1 orbit. Such an orbit would be represented in fig. 10 by a circle four yards in diameter.

All orbits, whether elliptic or circular, which have the same diameter, have also the same energy, but the energy changes when an electron crosses over from any orbit to another of a different diameter. Thus, to a certain limited extent, the atom constitutes a reservoir of energy. Its changes of energy are easily calculated; for example, the two orbits of smallest diameters in the hydrogen atom differ in energy by 16×10^{-12} ergs. If we pour radiation of the appropriate wave-length on to an atom in which the electron is describing the smallest orbit of all, it crosses over to the next orbit, absorbing 16×10^{-12} ergs of energy in the process, and so becoming temporarily a reservoir of energy holding 16×10^{-12} ergs. If the atom is in any way disturbed from outside, it may of course discharge the energy at any time, or it may absorb still more energy and so increase its store.

If we know all the orbits which are possible for an atom of any type, it is easy to calculate the changes of energy involved in the various transitions between them. As each transition absorbs or releases exactly one quantum of energy, we can immediately deduce the frequencies of the light emitted or absorbed in these transitions. In brief, given the arrangement of atomic orbits, we can calculate the spectrum of the atom. In practice the problem of course takes the converse form: given the spectrum, to find the structure of the atom which emits it. Bohr's model of the hydrogen atom is a good model at least to this extent—that the spectrum it would emit reproduces the hydrogen spectrum almost exactly. Yet the agreement is not quite perfect, and it is now generally accepted that Bohr's scheme of orbits is inadequate to account for actual spectra. We continue to discuss Bohr's scheme, not because the atom is actually built that way, but because it provides a good enough working model for our present purpose.

An essential, although at first sight somewhat unexpected, feature of the whole theory is that even if the hydrogen atom charged with its 16×10^{-12} ergs of energy is left entirely undisturbed, the electron must, after a certain time, lapse back spontaneously to its original smaller orbit, ejecting its 16×10^{-12} ergs of energy in the form of radiation in so doing. Einstein shewed that, if this were not so, then Planck's well-established "cavity-radiation" law could not be true. Thus a collection of hydrogen atoms in which the electrons describe orbits larger than the smallest possible orbit is similar to a collection of uranium or other radio-active atoms, in that the atoms spontaneously fall back to their states of lower energy as the result merely of the passage of time.

The electron orbits in more complicated atoms have much the same general arrangement as in the hydrogen atom, but are different in size. In the hydrogen atom the electron normally falls, after sufficient time, to the orbit of lowest energy and stays there. It might be thought by analogy that in more complicated atoms in which several electrons are describing orbits, all the electrons would in time fall into the orbit of lowest energy and stay there. Such does not prove to be the case. There is never room for more than one electron in the same orbit. This is a special aspect of a general principle which appears to dominate the whole of physics. It has a name—"the exclusion-principle"—but this is about all as yet; we have hardly begun to understand it. In another of its special aspects it becomes identical with the old familiar corner-stone of science which asserts that two different pieces of matter cannot occupy the same space at the same time. Without understanding the underlying principle, we can accept the fact that two electrons not only cannot occupy the same space, but cannot even occupy the same orbit. It is as though in some way the electron spread itself out so as to occupy the whole of its orbit, thus leaving room for no other. No doubt this must not be accepted as a literal picture of things, and yet it seems not improbable that the orbits of lowest energy in the hydrogen atom are possible orbits just because the electron can completely fill them, and that adjacent orbits are impossible because the electron would fill them $\frac{3}{4}$ or $1\frac{1}{2}$ times over, and similarly for more complicated atoms. In this connection it is perhaps significant that no single known phenomenon of physics makes it possible to say that at a given instant an electron is at such or such a point in an orbit of lowest energy; such a statement appears to be quite meaningless, and the condition of an atom is apparently specified with all possible precision by saying that at a given instant an electron is in such an orbit, as it would be, for instance, if the electron had spread itself out into a ring. We cannot say the same of other orbits. As we pass to orbits of higher energy, and so of greater diameter, the indeterminateness gradually assumes a different form, and finally becomes of but little importance. Whatever form the electron may assume while it is describing a little orbit near the nucleus, by the time it is describing a very big orbit far out it has become a plain material particle charged with electricity.

Thus, whatever the reason may be, electrons which are describing orbits in the same atom must all be in different orbits. The electrons in their orbits are like men on a ladder; just as no two men can stand on the same rung, so no two electrons can ever follow one another round in the same orbit. The neon atom, for instance, with 10 electrons, is in its normal state of lowest energy, when its 10 electrons each occupy one of the 10 orbits whose energy is lowest. For reasons which the quantum theory has at last succeeded in elucidating, there are, in every atom, two orbits in which the energy is equal and lower than in any other orbit. After this come eight orbits of equal but substantially higher energy, then 18 orbits of equal but still higher energy, and so on. As the electrons in each of these various groups of orbits all have equal energy, they are commonly spoken of, in a graphic but misleading phraseology, as rings of electrons. They are designated the K-ring, the L-ring, the M-ring, and so on. The K-ring, which is nearest to the nucleus, has room for two electrons only. Any further electrons are pushed out into the L-ring, which has room for eight electrons, all describing orbits which are different but of equal energy. If still more electrons remain to be accommodated they must go into the M-ring and so on.

In their normal states, the hydrogen atom has one electron in its K-ring, while the helium atom has two, the L, M, and higher rings being unoccupied. The atom of next higher complexity, the lithium atom, has three electrons, and as only two can be accommodated in its K-ring, one has to wander round in the outer spaces of the L-ring. In beryllium with four electrons, two are driven out into the *L*-ring. And so it goes on, until we reach neon with 10 electrons, by which time the *L*-ring as well as the inner *K*-ring is full up. In the next atom, sodium, one of the 11 electrons is driven out into the still more remote *M*-ring, and so on. Provided the electrons are not being excited by radiation or other stimulus, each atom sinks in time to a state in which its electrons are occupying its orbits of lowest energy, one in each.

So far as our experience goes, an atom, as soon as it reaches this state, becomes a true perpetual motion machine, the electrons continuing to move in their orbits (at any rate on Bohr's theory) without any of the energy of their motion being dissipated away, either in the form of radiation or otherwise. It seems astonishing and quite incomprehensible that an atom in such a state should not be able to yield up its energy still further, but, so far as our experience goes, it cannot. And this property, little though we understand it is, in the last resort, responsible for keeping the universe in being. If no restriction of this kind intervened, the whole material energy of the universe would disappear in the form of radiation in a few thousand-millionth parts of a second. If the normal hydrogen atom were capable of emitting radiation in the way demanded by the nineteenth century laws of physics, it would, as a direct consequence of this emission of radiation, begin to shrink at the rate of over a metre a second, the electron continually falling to orbits of lower and lower energy. After about a thousand-millionth part of a second the nucleus and the electron would run into one another, and the whole atom would probably disappear in a flash of radiation. By prohibiting any emission of radiation except by complete quanta, and by prohibiting any emission at all when there are no quanta available for dissipation, the quantum theory succeeds in keeping the universe in existence as a going concern.

It is difficult to form even the remotest conception of the realities underlying all these phenomena. The recent branch of physics known as "wave-mechanics" is at present groping after an understanding, but so far progress has been in the direction of co-ordinating observed phenomena rather than in getting down to realities. Indeed it may be doubted whether we shall ever properly understand the realities ultimately involved; they may well be so fundamental as to be beyond the grasp of the human mind.

It is just for this reason that modern theoretical physics is so difficult to explain, and so difficult to understand. It is easy to explain the motion of the earth round the sun in the solar system. We see the sun in the sky; we feel the earth under our feet, and the concept of motion is familiar to us from everyday experience. How different when we try to explain the analogous motion of the electron round the proton in the hydrogen atom! Neither you nor I have any direct experience of either electrons or protons, and no one has so far any inkling of what they are really like. So we agree to make a sort of model in which the electron and proton are represented by the simplest things known to us, tiny hard spheres. The model works well for a time and then suddenly breaks in our hands. In the new light of the wave-mechanics, the hard sphere is seen to be hopelessly inadequate to represent the electron. A hard sphere has always a definite position in space; the electron apparently has not. A hard sphere takes up a very definite amount of room, an electron—well, it is probably as meaningless to discuss how much room an electron takes up as it is to discuss how much room a fear, an anxiety or an uncertainty takes up, but if we are pressed to say how much room an electron takes up, perhaps the best answer is that it takes up the whole of space. A hard sphere moves from one point to the next; our model electron, jumping from orbit to orbit in the model hydrogen atom certainly does not behave like any hard sphere of our waking experience, and the real electron—if there is any such thing as a real electron in an atom—probably even less. Yet as our minds have so far failed to conceive any better picture of the atom than this very imperfect model, we can only proceed by describing phenomena in terms of it.

The Mechanical Effects of Radiation

The more compact an electrical structure is, the greater the amount of energy necessary to disturb it; and, as this energy must be supplied in the form of a single quantum, the greater the energy of the quantum must be, and so the shorter the wave-length of the radiation. A very compact structure can only be disturbed by radiation of very short wave-length.

A ship heading into a rough sea runs most risk of damage, and its passengers most risk of discomfort, when its length is about equal to the length of the waves. Short waves disturb a short ship and long waves a long ship, but a long swell does little harm to either. But this provides no real analogy with the effects of radiation, since the wave-length of radiation which breaks up an electrical structure is hundreds of times the size of the structure. The nautical analogy to such radiation is a very long swell indeed. As a rough working guide we may say that an electrical structure will only be disturbed by radiation whose wave-length is about equal to 860 times the dimensions of the structure, and will only be broken up by radiation whose wave-length is below this limit.* In brief, the reason why blue light affects photographic plates, while red light does not, is that the wave-length of blue light is less, and that of red light is greater, than 860 times the diameter of the molecule of silver chloride; we must get below the "860-limit" before anything begins to happen.

When an atom discharges its reservoir of stored energy, the light it emits has necessarily the same wave-length as the light which it absorbed in originally storing up this energy; the two quanta of energy being equal, their wave-lengths are the same. It follows that the light emitted by any electrical structure will also have a wave-length of about 860 times the dimensions of the structure. Ordinary visible light is emitted mainly by atoms, and so has a wave-length equal to about 860 atomic diameters. Indeed it is just because it has this wave-length that the light acts on the atoms of our retina, and so is visible.

Radiation of this wave-length disturbs only the outermost electrons in an atom, but radiation of much shorter wave-length may have much more devastating effects; X-radiation, for instance, may break up the far more compact inner rings of electrons, the K-ring, L-ring, etc., of the atomic structure. Radiation of still shorter wave-length may even disturb the protons and electrons of the nucleus. For the nuclei, like the atoms themselves, are structures of positive and negative electrical charges, and so must behave similarly with respect

* The mathematician will readily see the reason for this rule, which is, in brief, as follows: the energy needed to separate two electric charges $+e$ and $-e$, at a distance r apart, is e^2/r, and the energy needed to rearrange or break up a structure of electrons and protons of linear dimensions r will generally be comparable with this. If λ is the wave-length of the requisite radiation, the energy made available by the absorption of this radiation is the quantum hC/λ. Combining this with the circumstance that the value of h is very approximately 860 e^2/C, we find that the requisite wave-length of radiation is about 860 times the dimensions of the structure to be broken up.

to the radiation falling upon them, except for the wide difference in the wave-length of the radiation. Ellis and others have found that the γ-radiation emitted during the disintegration of the atoms of the radio-active element radium-B has wave-lengths of 3.52, 4.20, 4.80, 5.13, and 23×10^{-10} cms. These wave-lengths are only about a hundred-thousandth part of those of visible light, the reason being that the atomic nucleus has only about a hundred-thousandth part the dimensions of the complete atom. Radiation of such wave-lengths ought to be just as effective in re-arranging the nucleus of radium-B as that of 100,000 times longer wave-length is effective in re-arranging the hydrogen atom.

Since the wave-length of the radiation absorbed or emitted by an atom is inversely proportional to the quantum of energy, the quantum needed to "work" the atomic nucleus must have something like 100,000 times the energy of that needed to "work" the atom. If the hydrogen atom is a penny-in-the-slot machine, nothing less than five-hundred-pound notes will work the nuclei of the radio-active atoms.

The radio-active nuclei, like those of nitrogen and oxygen, could probably be broken up by a sufficiently intense bombardment, although the experimental evidence on this point is not very definite. If so, each bombarding particle would have to bring to the attack an energy of motion equal at least to that of one quantum of the radiation in question, this requiring it to move with an enormously high speed. Matter at sufficiently high temperatures contains an abundant supply both of quanta of high energy, and of particles moving with high speeds.

Temperature-Radiation. We speak in ordinary life of a red-heat or a white-heat, meaning the heat to which a substance must be raised to emit red or white light respectively.

The filament in a carbon-filament lamp is said to be raised to a red-heat, that in a gas-filled lamp to a yellow-heat. It is not necessary to specify the substance we are dealing with; if carbon emits a red light at a temperature of 3000°, then tungsten or any other substance, raised to this same temperature, will emit exactly the same red light as the carbon, and the same is true for other colours of radiation. Thus each colour, and so also each wave-length of radiation, has a definite temperature associated with it, this being the temperature at which this particular colour is most abundant in the spectroscopic analysis of the light emitted by a hot body. As soon as this particular temperature begins to be approached, but not before, radiation of the wave-length in question becomes plentiful; at temperatures well below this it is quite inappreciable.*

Just as we speak of a red-heat or a white-heat, we might, although we do not do so, quite legitimately speak of an X-ray heat or a γ-ray heat. The shorter the wave-length of the radiation, the higher the temperature especially associated with it. Thus as we make a substance hotter and hotter, it emits light of ever shorter wave-length, and runs in succession through the whole rainbow of colours—red, orange, yellow, green, blue, indigo, violet. We cannot command a sufficient range of temperature to perform the complete experiment in the laboratory, but nature performs it for us in the stars.

The Effects of Heat. We have already seen that radiation of short wave-length is needed to break up an electric structure of small dimensions. As short wave-lengths are asso-

---

* The wave-length λ of the radiation and the associated temperature T (measured in Centigrade degrees absolute) are connected through the well-known relation:
$$\lambda T = 0.2885 \text{ cm. degree.}$$

ciated with high temperatures, it now appears that the smaller an electrical structure is, the greater the heat needed to break it up. And we can calculate the temperature at which an electric structure of given dimensions will first begin to break up under the influence of heat.*

For instance, an ordinary atom with a diameter of about 4×10^{-8} cms. will first be broken up at temperatures of the order of thousands of degrees. To take a definite example, yellow light of wave-length 0.00006 cms. is specially associated with the temperature 4800 degrees; this temperature represents an average "yellow-heat." At temperatures well below this, yellow light only occurs when it is artificially created. But stars, and all other bodies, at a temperature of 4800 degrees emit yellow light naturally, and show lines in the yellow region of their spectrum, because yellow light removes the outermost electron from the atoms of calcium and similar elements. The electrons in the calcium atom begin to be disturbed when a temperature of 4800 degrees begins to be approached, but not before. This temperature is not approached on earth (except in the electric arc and other artificial conditions), so that terrestrial calcium atoms are generally at rest in their states of lowest energy.

To take another instance, the shortest wave-length of radiation emitted in the transformation of uranium is about 0.5×10^{-10} cms., and this corresponds to the enormously high temperature of 5,800,000,000 degrees. When some such temperature begins to be approached, but not before, the constituents of the radio-active nuclei ought to begin to re-arrange themselves, just as the constituents of the

* On combining the relation just given between T and λ with that implied in the rough law of the "860 limit," we find that a structure whose dimensions are r cms. will begin to be broken up by temperature-radiation when the temperature first approaches $1/3000r$ degrees.

calcium atom do when a temperature of 4800 degrees is approached.* This of course explains why no temperature we can command on earth has any appreciable effect in expediting or inhibiting radio-active disintegration.

The table below shows the wave-length of the radiation necessary to effect various atomic transformations. The last two columns shew the corresponding temperatures, and the kind of place, so far as we know, where this temperature is to be found, these latter entries anticipating certain results which will be given in detail in Chapter V below (p. 269). In places where the temperature is far below that mentioned in the last column but one, the transformation in question cannot be affected by heat, and so can only occur spontaneously. Thus it is entirely a one-way process. The available radiation not being of sufficiently short wave-length to work the atomic slot-machine, the atoms absorb no energy from the surrounding radiation and so are continually slipping back into states of lower energy, if such exist.

Highly-Penetrating Radiation

The shortest wave-lengths we have so far had under discussion are those of the γ-rays, but the last line of the table refers to radiation with a wave-length of only about a four-hundredth part of that of the shortest of γ-rays.

Since 1902, various investigators, Rutherford, Cooke, McLennan, Burton, Kolhörster and Millikan in particular, have found that the earth's atmosphere is continually being traversed by radiation which has enormously higher pene-

---

* If we suppose that rearrangements of an electric structure can also be effected by bombarding it with material particles, the temperature at which bombardment by electrons, nuclei, or molecules first becomes effective is about the same as that at which radiation of the effective wave-length would first begin to be appreciable; the two processes begin at approximately the same temperature.

EXPLORING THE ATOM

The Mechanical Effects of Radiation

Wave-lengths (cms.)	Nature of Radiation	Effect on Atom	Temperature (degrees abs.)	Where found
7500×10^{-8} to 3750×10^{-8}	Visible light	Disturbs outermost electrons	$3,850°$ to $7,700°$	Stellar atmospheres
250×10^{-8} to 10^{-8}	X-rays	Disturbs inner electrons	$115,000°$ to $29,000,000°$	Stellar interiors
5×10^{-9} to 10^{-9}	Soft γ-rays	Strip off all or nearly all electrons	$58,000,000°$ to $290,000,000°$	Central regions of dense stars
4×10^{-10}	γ-rays of radium-B	Disturbs nuclear arrangement	$720,000,000°$	
5×10^{-11}	Shortest γ-rays	—	$5,800,000,000°$	
1.3×10^{-13}	Highly-penetrating radiation (?)	Annihilation or creation of proton and accompanying electron	$2,200,000,000,000°$	

trating power than any known γ-rays. By sending up balloons to great heights, Kolhörster, and later Millikan and Bowen, have shewn that the radiation is noticeably more intense at great heights, thus proving that it comes into the earth's atmosphere from outside. If the radiation had its origin in the sun and stars, the main part of the radiation received on earth would come from the sun, and the radiation would be more intense by day than by night. This is found not to be the case, so that the radiation cannot come from the stars, and so must originate in nebulae or cosmic masses other than stars. Millikan is confident that its sources lie outside the galactic system.

The amount of the radiation is very great. Even at sea-level, where it is least, Millikan and Cameron find that it breaks up about 1.4 atoms in every cubic centimetre of air each second. It must break up millions of atoms in each of our bodies every second and we do not know what its physiological effects may be. The total energy of the radiation received on earth is just about a tenth of that of the total radiation, light and heat together, received from all the stars. This does not mean that light and heat are ten times as abundant as this radiation in the universe as a whole. For if the radiation originates in extra-galactic regions, then the stars which send us light and heat are comparatively near, while the sources of the highly-penetrating radiation are far more remote. On taking an average through the whole of space, including the vast stretches of internebular space, it seems likely that the highly-penetrating radiation is far more plentiful than stellar light and heat, and so is the most abundant form of radiation in the whole universe.

It is the most penetrating form of radiation known. Ordinary light will hardly pass through metals or solid substances at all; only a tiny fraction emerges through the thinnest of gold-leaf. On account of their shorter wave-length, and so of their more energetic quanta, X-rays will pass through foils of a few millimetres thickness of gold or of lead. The most highly-penetrating γ-rays from radium-B will pass through inches of lead. The radiation we have just been discussing varies in penetrating power; the most penetrating part of it will pass through 16 feet of lead.

It is not altogether clear whether the radiation is of the nature of very short γ-radiation or is of a corpuscular nature, like β-radiation; it may even be a mixture of both. Its penetrating power far exceeds that of any known β-radiation, so that if it is corpuscular, the corpuscles, must be moving with very nearly the velocity of light.

If, as seems far more likely, the radiation is, in part at least, of the nature of γ-radiation, then it ought to be possible to determine its wave-length from its penetrating power. Until quite recently different theories on the relation between the two have been in the field. The latest theory of all, that of Klein and Nishina, which is more perfect and more complete than any of the earlier theories, assigns to the most penetrating part of the radiation the amazingly short wave-length of 1.3×10^{-13} cms., as indicated in the table on p. 135.

We perhaps get the clearest conception of what this means if we apply the 860-rule; this shews that the radiation would break up an electric structure whose dimensions are only about 10^{-10} cms. No structure formed of electrons and protons can possibly be as small as this, for the radius of a single electron is about 2×10^{-13} cms. The radiation is of about the wave-length needed to break up the proton itself, the smallest and most compact structure known to science.

Approaching the problem from another angle, the

numerical relations already given shew that a quantum of radiation of this wave-length must have energy equal to 0.0015 erg, and so must have a weight of 1.7×10^{-24} grammes. Every physicist recognises this weight at once, for the best determinations give the weight of the hydrogen atom as 1.662×10^{-24} grammes. The quantum of highly-penetrating radiation has, then, just about the weight, and just about the energy, that would result from a complete hydrogen atom suddenly being annihilated and having all its energy set free as radiation.

It can hardly be supposed that all the highly-penetrating radiation received on earth has its origin in the annihilation of hydrogen atoms. If for no other reason, there are probably not enough hydrogen atoms in the universe for such a hypothesis to be tenable. The hydrogen atom consists of a proton and an electron, and its weight is roughly the same as the combined weight of a proton and an electron selected from any atom in the universe, so that, to a near enough approximation, the quantum of highly-penetrating radiation has the wave-length and energy which would result from a proton and electron in any atom whatever coalescing and annihilating one another. We have seen how the weights of the different known types of atoms approximate to integral multiples of the weight of the hydrogen atom, or to be more precise, differ by almost exactly equal steps, each of which is about equal to the weight of the hydrogen atom. The weight of the quantum of highly-penetrating radiation is equal to the change of weight represented by a single step, so that the quantum could be produced by any transformation which degraded the weight of an atom by a single step. In the most general case possible, this degradation of weight must, so far as we can see, arise from the coalescence of a proton and electron, with the resulting annihilation of both.

While this seems far and away the most probable source of this radiation, it is not the only conceivable source. For instance, the most abundant isotope of mercury, of atomic number 80 and atomic weight 200.016 is built of 200 protons, 120 nuclear electrons and 80 orbital electrons. Rutherford has pointed out that the sudden building up of such an atom out of 200 protons and 200 electrons would involve a loss of weight about equal to that of 1.5 atoms of hydrogen. If the building took place absolutely simultaneously, so that the whole of the liberated energy was emitted catastrophically as a single quantum, this quantum would have even more energy, and so be of even shorter wave-length, than the observed highly-penetrating radiation. Millikan had previously suggested the formation of other complex atoms out of simpler constituents as a possible source of the radiation, although it now appears that the schemes he propounded would not result in radiation of sufficiently short wavelength, at any rate if the modern Klein-Nishina theory is correct.

On the physical evidence alone, such schemes cannot be dismissed as impossible, but they must be treated as suspect on account of their high improbability. The mercury atom with its 400 constituent parts is a highly complicated structure, and it is exceedingly hard to believe that all these 400 parts could be hammered into a fully-formed atom by a single instantaneous act, accompanied by the catastrophic emission of only one quantum of radiation. If atoms ever are built up out of simpler constituents—and there is no evidence whatever that this process ever occurs in nature—it seems so much more likely that the aggregation would take place by distinct stages, and that the radiation would be emitted in a number of small quanta rather than in one large quantum. Moreover, any such hypothesis does not account for the numerical agreement of the calculated weight of the observed quanta of radiation with the known weight of the hydrogen atom. For these reasons, and on the general principle that the simpler and more natural hypothesis is always to be given preference in science, we may say that the annihilation of electrons and protons forms a more probable and more acceptable origin for the observed highly-penetrating radiation.

We may leave the problem in this state of uncertainty for the present, because it will appear later that astronomy has some evidence to give on the question.

CHAPTER III

Exploring in Time

WE have explored space to the furthest depths to which our telescopes can probe; we have explored into the intricacies of the minute structures we call atoms, of which the whole material universe is built; we now wish to go exploring in time. Man's individual span of life, and indeed the whole span of time covered by our historical records—some few thousands of years at most—are both far too short to be of any service for our purpose. We must find far longer measuring rods with which to sound the depths of past time and to probe forward into the future.

Our general method will be one which the study of geology has already made familiar. Undeterred by the absence of direct historical evidence, the geologist insists that life has existed on earth for millions of years, because fossil remains of life are found to occur under deposits which, he estimates, must have taken millions of years to accumulate. As he digs down through different strata in succession, he is exploring in time just as truly as the geographer who travels over the surface of the earth is exploring in space. A similar method can be used by the astronomer. We find some astronomical effect, quality, or property, which exhibits a continual accumulation or decrease, like the sand in the bottom or top half of the hour-glass; we estimate the rate at which this increase or decrease is occurring at the present moment, and also, if we can, the rate at which it must have occurred under the different conditions prevailing in the past. It then becomes a question, perhaps of mere arithmetic, although possibly of more complicated mathematics, to estimate the time which has elapsed since the process first started.

The Age of the Earth

The method is well exemplified in the comparatively simple problem of the age of the earth.

The first scientific attempt to fix the age of the earth was made by Halley, the astronomer, in the year 1715. Each day the rivers carry a certain amount of water down to the sea, and this contains small amounts of salt in solution. The water evaporates and in due course returns to the rivers; the salt does not. As a consequence the amount of salt in the oceans goes on increasing; each day they contain a little more salt than they did on the preceding day, and the present salinity of the oceans gives an indication of the length of time during which the salt has been accumulating. "We are thus furnished with an argument," said Halley, somewhat optimistically, "for estimating the duration of all things."

This line of argument does not lead to very precise estimates of the earth's age, but calculations based on modern data suggest that it must be many hundreds of millions of years.

More valuable information can be obtained from the accumulation of sediment washed down by the rain. Every year that passes witnesses a levelling of the earth's surface. Soil which was high up on the slopes of hills and mountains last year has by now been washed down to the bottoms of muddy rivers by the rain and is continually being carried out to sea. The Thames alone carries between one and two million tons of soil out to sea every year. For how long will England last at this rate, and for how long can it have already lasted? In our own lifetimes we have seen large masses of land round our coasts form landslides, and either fall wholly into the sea or slip down nearer to sea-level. Such conspicuous land-marks as the Needles, and indeed a large part of the southern coast of the Isle of Wight, are disappearing before our eyes. The geologist can form an estimate of the rapidity with which these and similar processes are happening, and so can estimate how long sedimentation has been in progress to produce the observed thickness of geological layers.

These thicknesses are very great; Professor Arthur Holmes [*] gives the observed maximum thicknesses as follows:

Pre-Cambrian	at least 180,000 feet
Palaeozoic Era (Ancient life)	185,000 "
Mesozoic Era (Mediaeval life)	91,000 "
Cainozoic Era (Modern life)	73,000 "

We can form a general idea of the rate at which these sediments have been deposited. Since Rameses II reigned in Egypt over 3000 years ago, sediment has been deposited at Memphis at the rate of a foot every 400 or 500 years; the excavator must dig down 6 or 7 feet to reach the surface of Egypt as it stood when Rameses II was king. The present rate of denudation in North America is estimated to be one foot in 8600 years; similar estimates for Great Britain indicate a rate of one foot in 3000 years. With geological strata deposited at an average rate of one foot per 1000 years, the total 529,000 feet of strata listed above would require over 500 million years for their deposition. At a rate of one foot

[*] In discussing the earth's age, I have borrowed extensively from Professor Holmes' book, *The Age of the Earth.*

per 4000 years, the time would be about 2100 million years.

This method of estimating geological times has been described as the "Geological hour-glass." We see how much sand has already run, we notice how fast it is running now, and a calculation tells us how long it is since it first started to run. The method suffers from the usual defect of hour-glasses, that there is no guarantee that the sand has always run at a uniform rate. Geological methods suffice to shew that the earth must be hundreds of millions of years old, but to obtain more definite estimates of its age, the more precise methods of physics and astronomy must be called in. Fortunately the radio-active atoms discussed in the previous chapter provide a perfect system of clocks, whose rate so far as we know does not vary by a hair's breadth from one age to another.

We have seen how, with the lapse of sufficient time, an ounce of uranium disintegrates into 0.865 ounces of lead and 0.135 ounces of helium. The process of disintegration is absolutely spontaneous; no physical agency known in the whole universe can either inhibit or expedite it in the tiniest degree. The following table shews the rate at which it progresses:

History of One Ounce of Uranium

Initially:		1 oz. uranium	No lead
After 100 million years	0.985 oz. uranium	0.013 oz. lead
" 1000 " "	0.865 " "	0.116 " "
" 2000 " "	0.747 " "	0.219 " "
" 3000 " "	0.646 " "	0.306 " "

and so on. Thus a small amount of uranium provides a perfect clock, provided we are able to measure the amount of lead it has formed, and also the amount of uranium still surviving, at any time we please. When the earth first solid-

EXPLORING IN TIME

ified, many fragments of uranium were imprisoned in its rocks, and may now be used to disclose the age of the earth. We are not entitled to assume that all the lead which is found associated with uranium has been formed by radio-active integration. But, by a fortunate chance, lead which has been formed by the disintegration of uranium is just a bit different from ordinary lead; the latter has an atomic weight of 207.2, while the former is of atomic weight only 206.0. Thus a chemical analysis of any sample of radio-active rock shews exactly how much of the lead present is ordinary leads and how much has been formed by radio-active disintegration. The proportion of the amount of lead of this latter kind to the amount of uranium still surviving tells us exactly for how long the process of disintegration has been going on.

In general all the samples of rock which are examined tell much the same story, and the radio-active clock is found to fix the time since the earth solidified at 1400 million years or more. The clock cannot tell us for how long before this the earth had existed in a plastic or fluid state, since in this earlier state the products of disintegration were liable to become separated from one another.

Aston has recently discovered a new isotope (see p. 111) of uranium, called actino-uranium. As uranium and its isotope have different periods of decay, the relative abundance of the two is continually changing. From the ratio of the amounts of these substances now surviving on earth, Rutherford has calculated that the age of the earth cannot exceed 3400 million years, and is probably substantially less.

These two physical estimates of the time which has elapsed since the earth solidified stand as follows:

Age of the Earth by the Radio-active Clock

1. From the lead-uranium ratio in radio-active rocks } More than 1400 million years.
2. From the relative abundance of uranium and actino-uranium } Less than 3400 million years.

Various astronomical methods are also available for determining the time since the solar system came into being. Here the "clocks" are provided by the shapes of the orbits of various planets and satellites. The orbits do not change at uniform rates, but their changes are determined by known laws, so that the mathematician can calculate the rates at which change occurred under past conditions, and hence, by totalling up, can deduce the time needed to establish present conditions. The following two estimates are both due to Dr. H. Jeffreys:

Age of the Solar System by the Astronomical Clock

1. From the orbit of Mercury...From 1000 to 10,000 million years.
2. "   "   the Moon...Roughly about 4000 million years.

While these various figures do not admit of any very exact estimate of the earth's age, they all indicate that this must be measured in thousands of millions of years. If we wish to fix our thoughts on a round number, probably 2000 million years is the best to select.

THE AGES OF THE STARS

We now turn to the far more difficult problem of determining the ages of the stars.

We shall not approach it by a direct frontal attack, but start far away from our real objective. Let us in fact start at the extreme other end of the universe, and delve a bit further into the properties of a gas.

Equipartition of Energy in a Gas. We have pictured a

gas as an indiscriminate flight of molecule-bullets. These fly equally in all directions, occasionally crashing into one another, and in so doing, changing both their speeds and directions of flight. We have seen that the total energy of motion undergoes no decrease when such collisions occur. If one of the molecules taking part in a collision has its speed checked, the other has its speed increased by such an amount that the energy lost by one molecule is gained by the other. Total energy of motion is "conserved."

Into this random hail of bullets, let us imagine that we project a far heavier projectile, which we may call a cannon-ball, with a speed equal to about the average speed of the bullets. The energies of the various projectiles are proportional jointly to their weights, and the squares of their speeds, so that in the present case, in which the speeds are all much the same, the big projectile has more energy than the bullets simply on account of its greater weight. If it weighs as much as a thousand bullets, it has a thousand times as much energy as each single bullet.

Yet the heavy projectile cannot for long continue swaggering through its lesser companions with a thousand times its fair share of energy. Its first experience is to encounter a hail of bullets on its chest. Very few bullets hit it in the back, for they are only moving at about its own speed, and so can hardly overtake it from behind. Moreover, even if they do, their blows on its back are very feeble because they are hardly moving faster than it. But the shower of blows on its chest are serious; every one of these tends to check its speed, and so to lessen its energy. And as the total energy of motion is conserved at every collision, it follows that, while the big projectile is losing energy all the time, the little ones must be gaining energy at its expense.

For how long will this interchange of energy go on? Will it, for instance, continue until the big projectile has lost all its energy, and been brought completely to rest? The problem is one for the mathematician, and it admits of a perfectly exact mathematical solution, which Maxwell gave as far back as 1859. The big projectile is not deprived of all its energy. As its speed gradually decreases, conditions change in all sorts of ways. When we allow for this change of conditions, we find that the energy of the big projectile goes on decreasing, not until it has lost all its energy, but until it has no more energy than the average bullet. When this stage is reached, the hits of the bullets are as likely on the average to increase the energy of the big projectile as to decrease it, so that this ends up by fluctuating around an amount equal to the average energy of the little projectiles.

Maxwell, and others after him, further shewed that no matter how many kinds of molecules there may be mixed together in a gas, and no matter how widely their weights may differ from one another, their repeated collisions must ultimately establish a state of things in which big molecules and little, light and heavy, all have the same average energy. This is known as the theorem of equipartition of energy. It does not mean that at any single instant all the molecules have precisely the same energy; obviously such a state of things could not continue for a moment, since the first collision between any pair of molecules would upset it immediately. But on averaging the energy of each molecule over a sufficiently long period of time—say a second, which is a very long time indeed in the life of a molecule, being the time in which at least a hundred million collisions occur—we shall find that the average energy of all the molecules is the same, regardless of their weights.

The same theorem can be stated in a slightly different form. Air consists of a mixture of molecules of different kinds and of different weights—molecules of helium which are very light, molecules of nitrogen which are far heavier, each weighing as much as seven molecules of helium, and the still heavier molecules of oxygen, each with the weight of eight molecules of helium. In its alternative form, the theorem tells us that at any instant the average energy of all the molecules of helium, in spite of their light weights, is exactly equal to the average energy of the molecules of nitrogen, and again each of these is exactly equal to the average energy of the molecules of oxygen. The lighter types of molecule make up for their small weights by their high speeds of motion. Similar statements are of course true for any other mixture of gases.

The truth of the theorem is confirmed observationally in a great variety of ways. In 1846, Graham measured the relative speeds with which the molecules of different kinds of gas moved, by observing the rates at which they streamed through an orifice into a vacuum; these proved to be such that the average energies of the various types of molecules were precisely equal to one another. Even earlier than this, Leslie and others had used this method to determine the relative weights of different molecules, although without fully understanding the underlying theory. Thus it may be accepted as a well-established law of nature that no molecule is allowed permanently to retain more energy than his fellows; in respect of their energies of motion, a gas forms a perfectly organised communistic state in which a law, which they cannot evade, compels the molecules to share their energies equally and fairly.

Subject to certain slight modifications, the same law applies also to liquids and solids. In liquids and gases, we can actually perform an experiment analogous to that of projecting our imaginary cannon-ball into the hail of molecule-bullets, and watch events. We may take a few grains of very fine powder, powdered gamboge or lycopodium seed, for instance, and let these play the part of super-molecules amongst the ordinary molecules of a gas or liquid. A powerful microscope shews that these super-molecules are not brought completely to rest, but retain a certain liveliness of movement, as they are continually hit about by the smaller and quite invisible true molecules. It looks for all the world as though they were affected by a chronic St. Vitus' dance, which shews no signs of diminishing as time goes on. These movements are called "Brownian movements," after Robert Brown, the botanist, who first observed them in the sap of plants. Brown at first interpreted them as evidence of real life in the small particles affected by them, an interpretation which he had to abandon when he found that particles of wax shewed the same movements. In a series of experiments of amazing delicacy, Perrin not only observed, but also measured, the Brownian movements of small solid particles as they were hit about by the molecules of air and other gases, and deduced the weights of the molecules of these gases with great accuracy.

Stellar Equipartition of Energy. We can now get back to the stars. The theorem of equipartition of energy is true not only of the molecules of a gas, and of a solid, and of a liquid; it is true also of the stars of the sky. The processes of mathematics are applicable to the very great as well as to the very small, and a theorem which is proved true for the minutest of atoms is equally true for the most stupendous of stars, provided of course that the premises on which it is based remain true, and do not suffer by transference from the small to the great end of the universe.

Now the conditions which are necessary for the theorem of equipartition of energy to be true happen to be amazingly simple; indeed it is difficult to believe that such wide consequences can follow from such simple conditions. They amount to practically nothing beyond a law of continuity and a law of causation; in other words, that the state of the system at any instant shall follow inevitably from its state at the preceding instant, or if you like, that there shall be no free-will among the molecules or stars or other bodies whose motions are under discussion. In the present turmoil as to the fundamental laws of physics, we cannot be entirely certain as to how far these very simple conditions are fulfilled in the molecular problem, although abundant observational evidence makes it clear that the law of equipartition holds, at any rate to an exceedingly good approximation, in an ordinary gas.

On the other hand, there is not the slightest doubt as to what determines the motions of the stars; it is the law of gravitation, every star attracting every other star with a force which varies inversely as the square of their distance apart. This is Newton's form of the law, but it is a matter of complete indifference for our present purpose whether we use the law in Newton's or in Einstein's form; for stellar problems the two are practically indistinguishable, and there is abundant evidence, particularly from the observed orbits of binary stars, in favour of either. The essential point is that, from the single supposition that the motions of the stars are governed by either of these laws of gravitation—or, for the matter of that, by any other not entirely dissimilar law—we can prove the theorem of equipartition of energy to be true for these motions. No subtle statement of exact conditions is required; the mere law of gravitation, together with the supposition that the stars cannot exercise free-will as to whether they obey it or not, are enough.

It is important to understand quite clearly what precisely the theorem asserts when applied to the stars. It does not of course assert that all the stars in the sky have equal energies. It does not even assert that on the average the heavy-weight stars in the sky have the same energy as the light-weight stars. What it asserts is that if we put any miscellaneous assortment of stars into space, then, after they have interacted with one another *for a sufficient length of time* (this is the essential point), those which started with more than their fair share of energy will have been compelled to hand over their excess to stars with lesser energy, so that the average energy of all different types of stars must necessarily become reduced to equality *in the long run*.

In the molecular problem, the interaction between the molecules takes place through the medium of collisions, and equipartition of energy is established, to a very good approximation, after some eight or ten collisions have happened to each molecule. In ordinary air, this requires a period of only about a hundred millionth part of a second.

In the stellar problem, we are dealing with very different lengths of time; collisions only occur at intervals of thousands of millions of millions of years. If the stars only redistributed their energy when actual collisions occurred, we might surmise that a close approximation to equipartition of energy would not be attained until after each star had experienced eight or ten collisions, and this would require a really stupendous length of time. Actually no such length of time is needed because the numerous gravitational pulls, even between stars which are at a considerable distance apart, equalise energy far more efficiently and expeditiously than the very rare direct hits. Every time that two stars happen to pass even fairly near to one another in their wanderings, each pulls the other a bit out of its course, and the directions and speeds of motion of both stars are changed—by much or little according as the stars pass quite close to one another or keep at a substantial distance apart. In brief, each approach of stars causes an interchange of energy, and after sufficient time, these repeated interchanges of energy result in the total energy being shared equally, on the average, between the stars, regardless of differences in their weights.

Now the crux of the situation, to which all this has been leading up, is that observation shews that stars of different weights are moving with different average speeds, these average speeds being such that equipartition of energy already prevails among the stars—not absolutely exactly, but to a tolerably good approximation.

The question of how long the stars must have interacted to reach such a condition now becomes one of absolutely fundamental importance, for *the answer tells us the ages of the stars*.

Stellar Velocities. We have already seen (p. 46) how stars which form binary systems can be weighed; such weighings disclosing weights ranging from about a hundred times the weight of the sun to only a fifth of its weight. The speeds of motion of binary systems can be measured in precisely the same way as the speeds of single stars. As far back as 1911, Halm, with an accumulation of such measurements before him, pointed out that the heaviest stars moved the most slowly. He found that, on the average, the heaviest of known stars had approximately the same energy of motion as the lightest, the high speeds of the latter just about making up for the smallness of their weights, and so suggested that the velocities of the stars, like those of the molecules of a gas, might be found to conform to the law of equipartition of energy. It appeared to be a case of Brownian movements on a stupendous scale.

Since then a great deal more observational evidence has accumulated, and an exhaustive investigation made by Dr. Seares of Mount Wilson in 1922 leaves very little room for doubt that the motions of the stars shew a real, and fairly close, approximation to equipartition of energy. The following table shews the final results of Seares' discussion.

The stars are first classified according to the different types of spectrum their light shews when analysed in a spectroscope.

Equipartition of Energy in Stellar Motions

Type of star	Average weight M (grammes)	Average speed C (cms. a sec.)	Average energy $\tfrac{1}{2}MC^2$ (ergs)	Corresponding temperature (degrees)
Spectral type $B\,3$	19.8×10^{33}	14.8×10^5	1.95×10^{46}	1.0×10^{62}
" $B\,8.5$	12.9	15.8	1.62	0.8
" $A\,0$	12.1	24.5	3.63	1.8
" $A\,2$	10.0	27.2	3.72	1.8
" $A\,5$	8.0	29.9	3.55	1.7
" $F\,0$	5.0	35.9	3.24	1.6
" $F\,5$	3.1	47.9	3.55	1.7
" $G\,0$	2.0	64.6	4.07	2.0
" $G\,5$	1.5	77.6	4.57	2.2
" $K\,0$	1.4	79.4	4.27	2.1
" $K\,5$	1.2	74.1	3.39	1.7
" $M\,0$	1.2	77.6	3.55	1.7

These different types of stars have very different average weights; the second column of the table shews that they exhibit a range of over 16 to 1. The third column, which gives the average speeds of these different types of stars, shews that the heaviest stars move the most slowly, and the lightest on the whole the most rapidly. The next column gives the average energy of motion of the different types of stars. This shews that the variation in speeds is just about that needed to make the average energies of all types of stars equal. An exception certainly occurs in the first two lines, which refer to the heaviest stars of all. Apart from these, the remaining ten lines shew a ratio of 10 to 1 in weight, whereas the average deviation of energy from the mean is only one of 9 per cent.

From this we see that the motions of the stars shew a real approach, and even a fairly close approach, to equipartition of energy. The question which naturally presents itself is whether this approximate equality of energy can be attributed to any other cause than long-continued gravitational interaction between the stars. This latter agency could undoubtedly produce it, but could anything else produce a similar result? The last column of the table provides the answer. It shews the temperatures to which a gas would have to be raised, in order that each of its molecules should have the same energy as the different types of stars. This may well seem an absurd calculation. A star weighing millions of millions of millions of tons goes hurtling through space at a speed of about 1,000,000 miles an hour; are we seriously setting out to inquire how hot a gas must be for every single one of its tiny molecules to have the same energy of motion, the same power of doing damage—for that is what energy of motion really amounts to—as the star? The calculation is undoubtedly absurd, and it is meant to be, because it is leading up to a *reductio ad absurdum*. If the observed equipartition of energy were brought about by any physical agency, such as pressure of radiation, bombardment by molecules, by atoms or by high speed electrons, this agency would have to be at a temperature, or in equilibrium with matter at a temperature, of the order of those given in the last column. These are temperatures of the order of 10^{6^2} degrees. We can be pretty sure no such temperature exists in nature, whence the argument runs that the observed equipartition of energy cannot have been brought about by physical means, and so must be the result of gravitational interaction between the stars.

The age of the stars is, then, simply the length of time needed for gravitational forces to bring about as good an approximation to equipartition of energy as is observed.

The calculation of this length of time presents a complicated but by no means intractable problem. All the necessary data are available, and as the method of calculation is well understood from previous experience in the theory of gases, the mathematician may be trusted to supply a reliable and reasonably exact answer when we ask him, but even without his help we can see that the time must be very long indeed.

Leaving actual figures aside for the moment, we may find it easier to think in terms of the scale-model we constructed in the first chapter (p. 82). We took our scale so small that the stars were reduced to tiny specks of dust; we noticed that space is so little crowded with stars that in our model the specks of dust had to be placed over 200 yards apart; to put it all in a concrete form, we found that Waterloo Station with only six specks of dust left in it is more crowded with dust than space is with stars. Now let the model come to life, so as to represent the motions of the stars. To keep the proportions right, the speed of the stars must of course be reduced in the same proportion as the linear dimensions of the model. In this the earth's yearly journey round the sun of 600 million miles had become reduced to a pin-head one-sixteenth of an inch diameter, say a fifth of an inch in circumference. As the stars move through space with roughly the same speed as the earth in its orbit, we may suppose the yearly journey of each speck of dust in our model also to be about a fifth of an inch. Thus each speck of dust will move about an inch in five years, roughly 16 feet in a thousand years—or say a ten-millionth part of a snail's pace. Even if two specks started moving directly towards one another it would take them about 20,000 years to meet. For how long must six particles of dust, floating blindly about in Waterloo Station, move at this pace before each has had enough close meetings with other specks of dust for their energy of motion to become thoroughly redistributed?

The mathematician, carrying out exact calculations with respect to the actual weights, speeds and distances of the stars, finds that the observed degree of approximation to equipartition of energy shews that gravitational interaction must have continued through millions of millions of years, most probably from 5 to 10 millions of millions of years. This, then, must be the length of life of the stars.

It is a stupendous length of time, and before finally accepting it we may well look for confirmation from other sources. In estimating the age of the earth we were able to invoke assistance from all kinds of clocks, astronomical, geological and physical; happily they all told much the same story. In the present problem only astronomical clocks are available, but fortunately there are no fewer than three of these, and again they agree in saying much the same thing.

The Orbits of Binary Systems. We have already seen

(p. 45) how the two constituents of a binary system permanently describe closed elliptical orbits about one another, because neither can escape from the gravitational hold of its companion. Energy can reside in the orbital motion of these systems, as well as in their motion through space. And strict mathematical analysis shews that a long succession of gravitational pulls from passing stars must finally result in equipartition of energy, not only between the energies of motion of one system and another through space, but also between the various orbital motions of which each binary system is capable. When this final state of equipartition is ultimately reached, the orbits of the systems will not all be similar, but it can be shewn that their shapes will be distributed according to a quite simple statistical law.* As the orbits of actual systems are not found to conform to this law, it is clear that the stars have not yet lived long enough to attain equipartition of energy in respect of their orbital motions. It is impossible to discuss how far they have travelled along the road to equipartition without knowing the point, or points, from which they started.

The question of the origin of binary systems will be discussed more fully in the next chapter. For the moment it may be said that they appear to come into being in two distinct ways.

Practically all astronomical bodies are in a state of rotation about an axis. The earth rotates about its axis once every 24 hours, and Jupiter once every 10 hours, as is shewn by the motion of the red spot and other markings on its surface. The surface of the sun rotates every 26 days or so; we can follow its rotation by watching sun-spots, faculae

* The eccentricity of orbit e is distributed in such a way that all values of e^2 from $e^2=0$ to $e^2=1$ are equally probable.

and other features moving round and round its equator. There are theoretical grounds for supposing that the sun's central core rotates considerably faster than this, most probably performing a complete rotation in comparatively few days. And it is likely that all the other stars in the sky are also in rotation, some fast and some slow. We shall see later how, with advancing age, a star is likely to shrink in size, and this shrinkage generally causes its speed of rotation to increase. Now mathematical theory shews that there is a critical speed of rotation which cannot be exceeded with safety. If the star rotates too fast for safety, it simply bursts into two, much as a rotating fly-wheel may burst if it is driven at too high a speed. It is in this way that one class of binary stars come into being. With a few exceptions this class is identical with the class of spectroscopic binaries described in Chapter I (p. 50); the two component stars are generally too close together to appear as distinct spots of light in the telescope, only spectroscopic evidence telling us that we are dealing with two distinct bodies.

Another class of binaries, the visual binaries, which appear quite definitely as pairs of spots of light in the telescope, probably have a different origin. We shall see later how the stars first come into being as condensations of nebulous gas, a whole shoal being born when a single great nebula breaks up. It must often happen that adjacent condensations are so near as to be unable to elude each other's gravitational grip. In time these shrink down into normal stars, while the gravitational forces remain just as powerful as before, and we are left with a pair of stars which must permanently journey through space in double harness, because they have not energy of motion enough ever to get clear of one another's gravitational hold. This mechanism produces a class of binaries which is precisely similar to that formed by the break-up of single stars, except for an enormous difference in scale. The distance between the two components of such a system must be comparable with the original distance between separate condensations in the primeval nebula out of which the stars were born, and so is enormously greater than the corresponding distance in spectroscopic binaries, which is comparable only with the diameter of an ordinary star which has broken into pieces. This explains why visual binaries appear as distinct pairs of spots of light, while spectroscopic binaries do not.

In the final state of equipartition of energy, the shapes of the orbits will, as we have seen, be distributed according to a definite statistical law. This law of distribution is the same for all sizes of orbit. On the other hand, the time needed for equipartition of energy to bring this law about is not the same for all sizes of orbit; it is far greater for the compact orbits of the spectroscopic binaries than for the more open orbits of the visual binaries. The reason for this is that changes in the shape of an orbit are caused merely by the *difference* of the gravitational pulls of a passing star on the two components of the binary. If the two components are very close together, the passing star exerts practically the same forces on both. These forces affect the motions of the two components in precisely the same way, with the result that the motion of the binary system as a whole through space is changed, but the shape of orbit remains unaltered. The passing star gets a grip on the motion of the binary as a whole, but none on the orbits of the components. On the other hand, when the components are far apart, the gravitational forces acting on the two may be widely different, so that a substantial change in the shape of the orbit may result, even if the encounter is not a very close one. In visual binaries, in which the components are usually hundreds of millions of miles apart, the time necessary to establish the final distribution of the "eccentricities," by which the shapes of elliptical orbits are measured, is once again found to be of the order of millions of millions of years, but it is something like a hundred times as great as this for the far more compact spectroscopic binaries.

The following table, compiled from material given by Dr. Aitken of Lick Observatory, shews the observed distribution of eccentricities in the orbits of those binaries for which accurate information is available:

The Approach to Equipartition of Energy in Binary Orbits

Eccentricity of Orbits	Observed number of spectroscopic binaries	Observed number of visual binaries	Number to be expected theoretically when the final state is attained
0 to 0.2	78	7	6
0.2 " 0.4	18	18	18
0.4 " 0.6	16	28	30
0.6 " 0.8	6	11	42
0.8 " 1.0	1	4	54

Let us look first at the spectroscopic binaries. In the observed orbits, we see that low eccentricities predominate, no fewer than 78 out of 119 having an eccentricity of less than one-fifth. In other words, most spectroscopic binaries have nearly circular orbits. Both theory and observation shew that when a star first divides up into a spectroscopic binary, the orbits of the two components must be nearly circular, so that the table of observed orbits provides very little evidence of any progressive change of shape in the orbits as a whole. In contrast to this, the last column of the table shews the proportion of orbits of different eccentricities which is to be expected when, if ever, equipartition of energy is finally attained. Here high eccentricities, representing very elongated orbits, predominate; only one orbit in twenty-five is so nearly circular as to have an eccentricity less than a fifth.

In general the observed numbers tabulated in the second column shew no resemblance at all to the theoretical numbers tabulated in the fourth column. In other words, the spectroscopic binaries shew no suggestion of any near approach to the final state, most of them retaining the low eccentricity of orbit with which they started life. We should naturally expect this, since we have seen that hundreds or even thousands of millions of millions of years would be needed for these orbits to attain a final state of equipartition, and the stars cannot be as old as this, for if they were, their motions through space ought to shew absolutely perfect equipartition, which they certainly do not.

Turning now to the third column, we see that the visual binaries shew a good approach to the theoretical final state up to an eccentricity of about 0.6, but not beyond. The deficiency of orbits of high eccentricity may mean that gravitational forces have not had sufficient time to produce the highest eccentricities of all, but part, and perhaps all, of it must be ascribed to the simple fact that orbits of high eccentricity are exceedingly difficult to detect observationally and to measure accurately.

Clearly, then, the study of orbital motions, like that of motions through space, points to gravitational action extending over millions of millions of years. In each case there is an exception to "prove the rule." In the case we have just considered it is provided by the spectroscopic binaries, which are so compact that their constituents can defy the pulling-apart action of gravitation; in the former case it was provided by the B-type stars, which are so massive, possibly also so young, that the gravitational forces from less weighty stars have not yet greatly affected their motion.

When these two lines of evidence are discussed in detail, they agree in suggesting that the general age of the stars is about that already stated, namely, from five to ten millions of millions of years.

Moving Clusters. A third line of evidence, which also tells much the same story, may be briefly mentioned. The conspicuous groups of bright stars in the sky, such as the Great Bear, the Pleiades and Orion's Belt, consist for the most part of exceptionally massive stars which move in regular orderly formation through a jumble of slighter stars, like a flight of swans through a confused crowd of rooks and starlings. Swans continually adjust their flight so as to preserve their formation. The stars cannot, so that their orderly formation must in time be broken by the gravitational pull of other stars. The lighter stars are naturally knocked out of formation first, while the most massive stars retain their formation longest. Observation suggests that this is what actually happens to a moving star-cluster; at any rate the stars which remain in formation generally have weights far above the average. And, as we can calculate the time necessary to knock out the lighter stars, we can at once deduce the ages of those which are left in.

The result of the calculation confirms those already mentioned, so that we find that the three available astronomical clocks all tell much the same time. They agree in indicating an age of the order of five to ten millions of millions of years for the stars as a whole.

Another line of investigation, to be mentioned later (p. 178) again points to a similar age.

It is perhaps a little surprising that this age should prove to be so much longer than the age of the earth, although there is of course no positive reason why the earth should not have been born during the last few moments of the lives of the stars. It is perhaps also a little surprising that it should prove to be much longer than the age suggested, very vaguely it is true, by de Sitter's cosmology. If we accept the apparent velocities of recession of the most distant nebulae as real, we find that some thousands of millions of years of motion at their present speeds would just about account for their present distances from us, so that a few thousands of millions of years ago, the nebulae must have been far more huddled together than they now are. This is of course very different from saying that the time which has elapsed since the creation of the nebulae can only be a few thousands of millions of years, yet we might reasonably have expected *à priori* that the two periods would be at least comparable.

To state the difficulty in a slightly different form, a period of a thousand million years seems to have made a great deal of difference to the earth, and a great deal of difference to the general arrangement in space of the great nebulae, so that it is odd that it should make so little difference to the stars that we need to postulate an age a thousand times as great before we can explain their present condition.

These considerations may seem to suggest that the estimate just made of stellar ages should be accepted with caution and perhaps even with suspicion. Yet if we reject it, so many facts of astronomy are left up in the air without any explanation, and so much of the fabric of astronomy is thrown into disorder (see p. 177, below), that we have little option but to accept it, and suppose that the stars have actually lived through times of the order of millions of millions of years.

The Sun's Radiation

During the whole of some such vast period of time, the sun has in all probability been pouring out light and heat at least as profusely as at present. Indeed a mass of evidence, to which we shall return later, shews that young stars emit more radiation than older stars, so that during most of its long life the sun must have been pouring out energy even more lavishly than now.

If our ancestors thought about the matter at all, they probably saw nothing remarkable in this profuse outpouring of light and heat, particularly as they had no conception of the stupendous length of time during which it had lasted. It was only in the middle of last century, when the principle of conservation of energy first began to be clearly understood, that the source of the sun's energy was seen to constitute a scientific puzzle of really first-class difficulty. The sun's radiation obviously represented a loss of energy to the sun, and, as the principle of conservation shewed that energy could not originate out of nothing, this energy necessarily came from some source or store adequate to supply vast outpourings of energy over a very long period of time. Where was such a store to be found?

The sun at present pours out radiation at such a rate that if the necessary energy were generated in a power-station outside the sun, this station would have to burn coal at the rate of many thousands of millions of millions of tons a second. There is of course no such power-station. The sun is entirely dependent on its own resources; it is a ship on an empty ocean. And if, like such a ship, the sun carried its own store of coal, or if, as Kant imagined, its whole substance were its store of coal, so that its light and heat came from its own combustion, the whole would be burnt into ashes and cinders in a few thousand years at most.

The history of science records one solitary attempt to explain the sun's energy as coming in from outside. We have seen how the energy of motion of a bullet is transformed into heat when the speed of the bullet is checked. An astronomical example of the same effect is provided by the familiar phenomenon of shooting-stars. These are bullet-like bodies which fall into the earth's atmosphere from outer space. So long as such a body is travelling through empty space, its fall towards the earth continually increases its speed, but, as it enters the earth's atmosphere, its speed is checked by air-resistance, and the energy of its motion is gradually transformed into heat. The shooting-star becomes first hot and then incandescent, emitting the bright light by which we recognise it. Finally, the heat completely vaporises it, and it disappears from sight, leaving only a momentary trail of luminous gas behind. The original energy of motion of the shooting-star has been transformed into light and heat—the light by which we see it, and the heat by which it is ultimately vaporised.

In 1849, Robert Mayer suggested that the energy which the sun emitted as radiation might accrue to it from a continuous fall of shooting-stars or similar bodies into the solar atmosphere. The suggestion is untenable, because a simple calculation shews that a mass of such bodies equal to the weight of the whole earth would hardly maintain the sun's radiation for a century, and that the infall needed to maintain the sun's radiation for 30 million years would double its weight. As it is quite impossible to admit that the sun's weight can be increasing at any such rate, Mayer's hypothesis has to be abandoned.

In 1853 Helmholtz put forward a very similar theory, the famous "contraction-hypothesis," according to which the sun's own shrinkage sets free the energy which ultimately appears as radiation. If the sun's radius shrinks by a mile, its outer atmosphere falls through a height of a mile and sets free as much energy in so doing as would be yielded up by an equal weight of shooting-stars falling through a mile and having their motion checked. On Helmholtz's theory, the different parts of the sun's own body performed the *rôles* which Mayer had allotted to shooting-stars falling in from outside; they performed these same parts again and again, until ultimately the sun had shrunk so far that it could shrink no further. Yet Helmholtz's theory, like that of Mayer, failed to survive the test of numerical computation. In 1862 Lord Kelvin calculated that the shrinkage of the sun to its present size could hardly have provided energy for more than about 50 million years of radiation in the past, whereas the geological evidence already noticed (p. 142) shews that the sun must have been shining for a period enormously longer than this.

To track down the actual source of the sun's energy with any hope of success, we must give up guessing, and approach the problem from a new angle. We have seen (p. 113) how radiation carries weight about with it, so that any body which is emitting radiation is necessarily losing weight; the radiation emitted by a searchlight of 50 horse-power would, we saw, carry away weight at the rate of about a twentieth of an ounce a century. Now each square inch of the sun's surface is in effect a searchlight of just about 50 horse-power, whence we conclude that weight is streaming away from every square inch of the sun's surface at the rate of about a twentieth of an ounce a century. Such a loss of weight seems small enough, until we multiply it by the total number of square inches which constitute the whole surface of the sun. It then appears that the sun as a whole is losing weight at the rate of rather over 4 million tons a second, or about 250 million tons a minute—something like 650 times the rate at which water is streaming over Niagara.

The Past Histories of the Sun and Stars. Let us carry on the multiplication. Two hundred and fifty million tons a minute is 360,000 million tons a day. Thus the sun must have weighed 360,000 million tons more than now at this time yesterday, and will weigh 360,000 millions tons less at this time to-morrow. And 360,000 million tons a day is 131 million million tons a year. We can dig as far into the past as we like in this way and can probe as far as we like into the future. But soon we encounter the usual trouble which besets all calculations of this kind—the sand does not always run through the hour-glass at the same rate. The rate at which the sun loses weight will not vary appreciably between to-day and to-morrow, or even over a century or a million years, but we must be on our guard against going too far. If the sun continued to radiate at precisely its present rate, a simple sum in division shews that it would last for just about 15 million million years, by which time its last ounce of weight would be disappearing. Incidentally this gives us a vivid conception of the enormous weight of the sun; it could go on pouring away its substance into space at 650 times the rate at which water is pouring over Niagara for 15 million million years before becoming exhausted.

Obviously, however, we cannot carry out our calculations in this simple light-hearted way; it would be absurd to suppose that the sun's last ton of substance will radiate energy at the same rate as his present stupendous mass of two thousand million million million million tons. A series of investigations which culminated in a paper published by Eddington in 1924, disclosed that, in a general sort of way, a star's luminosity depends mainly on its weight. The dependence is not very precise, and neither is it universal, but when we are told a star's weight we can say that its luminosity is likely, with a high degree of probability, to lie within certain fairly narrow limits. For instance most stars whose weight is nearly equal to the sun are found to have about the same luminosity as the sun. In general, as might be expected, stars of light weight radiate less than heavy stars, but also—and this could not have been foreseen—the differences in their radiations are far greater than the differences in their weights. The law which we have already noticed to hold for a few stars in the neighbourhood of the sun is true, although in a somewhat different sense, for the stars as a whole—the candle-power per ton is greatest in the heaviest stars. For example, the average star of half the weight of the sun does not radiate anything like half as much energy as the sun: the fraction is more like an eighth. This consideration extends the future life of the sun, and indeed of all the stars, almost indefinitely. A sort of parsimony seems to creep over the stars in their old age; so long as they have plenty of weight to squander, they squander it lavishly, but they contract their scale of expenditure when they have little left to spend. The sand runs slowly through the hour-glass when there is little left to run.

In the same way, the average star of double the sun's weight does not merely radiate twice as much energy as the sun; it radiates about eight times as much. We must keep this in view in estimating the past life of the sun; it shortens the sun's past life just as surely as the opposite effect lengthens its future life. Observation tells us at what rate the average star of any given weight spends its weight in the form of radiation, and, on the supposition that the sun has behaved like this typical average star at the corresponding stage of its own past history, we can draw up a table exhibiting its gradual change of weight as its life progressed. Selected entries from this table would read somewhat as follows:

2,000,000,000 years ago, the sun had	1.00013	times its present weight
1,000,000,000,000 " "	1.07	" "
2,000,000,000,000 " "	1.16	" "
5,700,000,000,000 " "	double	"
7,100,000,000,000 " "	4 times	"
7,400,000,000,000 " "	8 "	"
7,500,000,000,000 " "	20 "	"
7,600,000,000,000 " "	100 "	"

The first entry represents roughly the time since the earth was born. It shews that, during the whole existence of the earth, the sun's weight has changed by only an inappreciable fraction of the whole. Consequently, it seems likely, although naturally we cannot be certain, that when the earth was born the sun was much the same as it now is, and that it has been the same, in all essential respects, throughout the whole life of the earth.

To come to appreciably different conditions we have to go back to remote aeons far beyond the time of the earth's birth. We are free to do this, for we have seen that the

EXPLORING IN TIME

earth's whole life is only a moment in the lives of the stars. We have estimated the latter as being something of the order of 5 to 10 million million years, and it is only when we go back an appreciable fraction of these long periods that we find the sun's weight differing appreciably from its present weight. We have, for instance, to go back more than 5 million million years to find the sun with double its present weight. When we go back much further than this a new phenomenon appears; the weight of our hypothetical past sun begins to go up by leaps and bounds. In time it begins to double and more than double every 100,000 million years, and we cannot go back as far as 8 million million years without postulating a sun of quite impossibly high weight. The sun must, then, have been born some time within the last 8 million million years.

The exact figures of our table may be open to suspicion, but as a general fact of observation there is no doubt that very massive stars radiate away their energy, and therefore also their weight, with extraordinary rapidity. Indeed the process is so rapid that we may disregard all that part of a star's life in which it has more than about 10 times the weight of the sun—this is lived at lightning speed. Apart from all detailed calculations, this general principle fixes a definite limit to the ages, not only of the sun, but also of every other star. The upper limit to the age of the sun is certainly somewhere in the neighbourhood of 8 million million years.

This agrees well enough with the general age of from 5 to 10 million million years that other calculations have assigned to the stars in general. The calculations thus reinforce one another, and it looks as if at least two of the pieces of the puzzle were beginning to fit satisfactorily together.

If all the stars in the sky were similar to the sun, we might feel a good deal of confidence in the conclusions we have reached.

Unfortunately, difficulties emerge as soon as we discuss the ages of stars which at present have many times the weight of the sun. The table on p. 154 shews that a class of stars (spectral type $A\ 0$) of six times the weight of the sun have motions in space which conform well enough to the law of equipartition of energy. Unless this is a pure coincidence (and this is unlikely, in view of the fact that other groups of only slightly less weight conform equally well), we must assign an age of from 5 to 10 million million years to these very massive stars. Yet the average star of this weight is emitting about a hundred times as much radiation as the sun, which means that it is halving its weight every 150,000 million years. Clearly this process cannot have gone on for anything like 5 or 10 million million years.

Still more luminous stars present the problem in an even more acute form. The star S Doradus in the Lesser Magellanic Cloud is at present emitting 300,000 times as much radiation as the sun. Whereas the sun is pouring its weight out into space at the rate of 650 Niagaras, S Doradus is pouring it out at the rate of 200,000,000 Niagaras; every 50 million years it loses a weight equal to the total weight of the sun. It is obviously absurd to imagine that this star can have been losing weight at this rate for millions of millions of years.

For such a star as S Doradus only two alternatives seem open. Either it was created quite recently (on the astronomical time scale), and so is still at the very beginning of its prodigal youth, or else its loss of weight has in some way been inhibited through the greater part of its life. A good many arguments weigh against the hypothesis of recent creation. The star is a member of a star cloud in which we should naturally expect all the members to be of approximately equal age. It is in a region of space in which there are no indications that stars are still being born. And, even if we accept the hypothesis of recent creation for this particular star, we are still at a loss to explain how the other massive stars, which figure in the table on p. 154, can be old enough for equipartition of energy to have become already established.

For many reasons it seems preferable, and indeed almost inevitable, to suppose that these highly luminous and very weighty stars have in some way been saved from energetic radiation, with its consequential rapid wasting of weight, throughout the greater part of their lives. In brief, we suppose that they are cases of arrested development, whose weight and general appearance equally belie their true ages. Later (p. 302) we shall come upon a physical mechanism which explains very simply and naturally how this could happen.

If this hypothesis can be accepted, it clears up the whole situation. As soon as we accept it, we become free to assign any age we please to the stars, and naturally select that indicated by the law of equipartition of energy, at any rate for those classes of stars which are found to conform to this law.

The exceptionally luminous stars which we have just had under discussion are comparatively rare objects in the sky. The vast majority of stars have luminosities and weights comparable with, or distinctly less than, those of the sun, and for these the difficulty does not exist. Indeed the hypothesis of arrested development would break down under its own weight if we had to invoke its help for many stars; it is tenable just because we seldom need to use it. We may accept the table on p. 170 as giving the past history of the sun with tolerable accuracy, thus fixing its age at something under 8 million million years, and a generally similar table would apply to most of the stars in the sky.

The Source of Stellar Energy

The ages of 5 million million years or more which we have been led to assign to the stars imply that at birth the sun must have had at least double, and more probably several times, its present weight. For every ton which existed in the sun at its birth only a few hundred-weight remain to-day. The rest of the ton has been transformed into radiation and, streaming away into space, has left the sun for ever.

In the preceding chapter, we had occasion to discuss the transformation of weight into radiation which accompanies the spontaneous disintegration of radio-active atoms. The most energetic instance of this phenomenon known on earth is the transformation of uranium into lead, in which about one part in 4000 of the total weight is transformed into radiation. In the sun, the corresponding fraction may be half, or nine-tenths, or even 99 per cent., but, whatever it is, it certainly exceeds one part in 4000. Thus the process by which the sun generates its light and heat must involve a far more energetic transformation of material weight into radiation than any process known on earth.

Perrin and Eddington at one time suggested that this process may be the building up of complex atomic nuclei out of protons and electrons. The simplest, and most favourable example of this, which was especially considered by Eddington, is to be found in the building up of the helium nucleus. The constituents of a helium atom are precisely identical with those of four hydrogen atoms, namely, four electrons and four protons. If these constituents could be rearranged without any transformation of material weight into radiation, the helium atom would have precisely four times the weight of the hydrogen atom. In actual fact Aston finds that the ratio of weights is only 3.970. The difference between this and 4.000 must represent the weight of the radiation which goes off when, if ever, the helium atom is built up by the coalescence of four hydrogen atoms. The loss of weight, one part in 130, is very much greater than occurs in radio-active transformations, but even so it does not provide adequate lives for the stars. The transformation of a sun which originally consisted of pure hydrogen into one consisting wholly of helium would only provide radiation at the sun's present rate of radiation for about 100,000 million years, and the dynamical evidence of equipartition of energy, etc., as well as other evidence which we shall consider later (p. 178 below), demands far longer lives for the stars than this.

The Annihilation of Matter. Modern physics is only able to suggest one process capable of providing a sufficiently long life for a radiating star; it is the actual annihilation of matter. Various lines of evidence go to shew that the atoms in very massive stars are not, for the most part, fundamentally different from those in less massive stars. Thus the primary cause of the difference in weight between a heavy star and a light star is not a difference in the quality of the atoms; it is a difference in their number. A heavy star can only change into a light star through the actual disappearance of atoms; these must be annihilated, and their weight transformed into radiation.

I first drew attention in 1904 to the large amount of energy capable of being liberated by the annihilation of matter, positive and negative electric charges rushing together, annihilating one another and setting their energy loose in space as radiation. The next year Einstein's theory of relativity provided a means for calculating the amount of energy which would be produced by the annihilation of a given amount of matter; it shewed that energy is set free at the rate of 9×10^{20} ergs per gramme, regardless of the nature or condition of the substance which is annihilated. I subsequently calculated the length of lives which this source of energy permitted to the stars, but the calculated lives of millions of millions of years seemed greater than were needed by the astronomical evidence available at the time. Since then a continual accumulation of new evidence, particularly that discussed in the present chapter, has been seen to demand stellar lives of precisely these lengths, with the result that the majority of astronomers now regard annihilation of matter as the most probable source of stellar energy.

Other considerations in addition to those just mentioned point to the annihilation of matter as the fundamental process going on in the stars. If there were no annihilation of matter, a star could only change its weight by some small fraction of the whole, such, for example, as the one part in 4000 which accompanies radio-active disintegration, or as the one part in 130 which would result from the building up of helium atoms out of hydrogen. A star would retain its weight practically unaltered through its life. This would of course necessarily impose far shorter lives on the stars than we believe them to have had, for nothing can alter the fact that the sun loses 360,000 million tons of weight every day in radiation, so that if its weight cannot change much, it cannot have radiated for long.

We have seen that in the present universe, a star's luminosity depends mainly on its weight. If we imagine that the same condition of things has always prevailed, then stars which retained the same weight throughout their lives would have to retain approximately the same luminosity, at any rate until their capacity for radiation became exhausted. Otherwise, contrary to observation, we should find stars with weights equal to the sun having all possible degrees of luminosity. Thus if we discard the hypothesis of the annihilation of matter, it becomes necessary to imagine some controlling mechanism, of a kind which would compel stars having the weight of the sun always to radiate at about the same rate as the sun, at least until sheer exhaustion prevents them from radiating any more, and similarly for stars of all other weights.

There does not seem to be any general objection against supposing such a controlling mechanism to exist, and indeed such mechanisms have been advocated by Russell and Eddington. But when we consider such a mechanism in detail, we encounter various objections which we shall consider in the next Chapter V, p. 276, the principal of which is that, stars controlled by it would, so far as we can see, be in a highly explosive state. And immediately we abandon the hypothesis of such a controlling mechanism, the observed close dependence of luminosity on weight compels us to suppose that a star's weight decreases as its luminosity diminishes, which leads us back immediately to the annihilation of matter.

A further consideration which points in the same direction may be mentioned here. We have seen how the "candle-power per ton of weight" is greatest in the heavier stars. As an immediate consequence the loss of weight per ton is greatest in the heaviest stars. In the time in which a massive star loses a hundred-weight per ton, a star of light weight may lose only a few pounds per ton. The consequence is that the passage of time tends to equalise the weights of the stars. This principle no doubt explains in large part why the present stars shew no very great range of weight. It also leads to interesting consequences when applied to the two components of a binary system. It shews that as a binary system ages, its two components ought continually to become more nearly equal in weight. Thus the two components ought to differ less in weight in old binaries than in young.

This last conclusion can be tested observationally. Aitken finds that the ratio of weights of the two constituents of a binary increases from about 0.70 for young systems of large weight to 0.90 for older systems in which the constituents are about similar to the sun. The direction of change is that predicted by theory; the amount of change indicates a time-interval of 5.4 million million years between the two states concerned. This agrees well enough with our previous estimates of the sun's age—but it is far less reliable as it is based on somewhat slender evidence.

On the whole, in whatever direction we try to escape from the hypothesis of annihilation of matter, the alternative hypothesis we set up to explain the facts seem to lead us back in time to the annihilation of matter.

We must not overlook the revolutionary nature of the change which this hypothesis introduces into physical science. The two fundamental corner-stones of nineteenth century physics, the conservation of matter and the conservation of energy, are both abolished, or rather are replaced by the conservation of a single entity which may be matter and energy in turn. Matter and energy cease to be indestructible and become interchangeable, according to the fixed rate of exchange of 9×10^{20} ergs per gramme.

Yet, looked at from another angle, the hypothesis only carries physics one stage further along the road it has already trodden in the past. Heat, light, electricity have all in turn proved to be forms of energy; the annihilation hypothesis only proposes to add another to the list, so that matter itself also becomes a form of energy.

According to this hypothesis all the energy which makes life possible on earth, the light and heat which keep the earth warm and grow our food, and the stored up sunlight in the coal and wood we burn, if traced far enough back, are found to originate out of the annihilation of electrons and protons in the sun. The sun is destroying its substance in order that we may live, or, perhaps we should rather say, with the consequence that we are able to live. The atoms in the sun and stars are, in effect, bottles of energy, each capable of being broken and having its energy spilled throughout the universe in the form of light and heat. Most of the atoms with which the sun and stars started their lives have already met this fate; the remainder are doubtless destined to meet it in time. Scientific writers of half a century ago delighted in the picturesque description of coal as "bottled sunshine"; they asked us to think of the sunshine as being bottled up as it fell on the vegetation of the primaeval jungle, and stored for use in our fireplaces after millions of years. On the modern view we must think of it as re-bottled sunshine, or rather re-bottled energy. The first bottling took place millions of millions of years ago, before either sun or earth were in being, when the energy was first penned up in protons and electrons. Instead of thinking prosaically of our sun as a mere collection of atoms, let us think of it for a moment as a vast storehouse of bottles of energy which have already lain in storage for millions of millions of years. So enormous is the sun's supply of these bottles, and so great the amount of energy stored in each that, even after radiating light and heat for 7 or 8 million million years, it still has enough left to provide light and heat for millions of millions of years yet to come.

Two quantitative considerations may help to shew these processes in a clearer light. We have seen that the sun's present store of atoms would, at the present rate of breakage, last for 15 million million years. This means that every year only one atom in 15 million million is broken, a fraction which may seem absurdly small to produce the sun's vast continuous outpourings of energy. Let us, however, reflect that the energy which is continually pouring out of the sun's surface at the rate of about 50 horse-power per square inch is generated throughout the vast interior of the sun's body; the stream of energy which emerges from a square inch of surface is the concentration of all the energy generated in a cone of a square inch cross-section, but of 433,000 miles depth. Such a cone contains about 10^{33} atoms, and although only one in 15 million million is broken each year, there are still about two million million atoms destroyed each second.

Even so, the amount of energy set free by the annihilation of matter is rather surprising; it is of an entirely different order of magnitude from that made available by any other treatment. The combustion of a ton of the best coal in pure oxygen liberates about 5×10^{16} ergs of energy; the annihilation of a ton of coal liberates 9×10^{26} ergs, which is 18,000 million times as much. In the ordinary combustion of coal we are merely skimming off the topmost cream of the energy contained in the coal, with the consequence that 99.999999994 per cent. of the total weight remains behind in the form of smoke, cinders or ash. Annihilation leaves nothing behind; it is a combustion so complete that neither smoke, ash, nor cinders is left. If we on earth could burn our coal as completely as this, a single pound would keep the whole British nation going for a fortnight, domestic fires, factories, trains, power-stations, ships and all; a piece of coal smaller than a pea would take the *Mauretania* across the Atlantic and back.

Purely astronomical evidence has led to the conclusion that atoms are continually being annihilated in the sun and stars. Here we have a piece of the puzzle which fits perfectly on to those we tentatively fitted together in the last chapter. As we there saw, recent investigations in mathematical physics suggest that the highly-penetrating radiation received on earth has its origin in the annihilation of matter out in space. And the amount of this radiation received on earth is so great that we had to suppose the underlying annihilation of matter to be one of the fundamental processes of the universe; we now discover that it is in all probability the process which keeps the sun and stars shining and the universe alive.

Physical Interpretation. It is perhaps worth trying to probe still one stage further into the physical nature of this process of annihilation of matter, although it must be premised that what follows is speculative in the sense that no direct observational confirmation is at present available.

We saw (p. 127) how the electrodynamical theory current in the last century required that the nucleus and electron of the hydrogen atom should approach ever closer and closer to one another with the mere passage of time, until finally they rushed together and coalesced. When this happened, the negative charge of the electron and the positive charge of the nucleus would neutralise one another and their energy would go off in a flash of radiation similar to the flash of lightning which indicates that the negative and positive charges in two opposing thunder-clouds have met and neutralised one another.

The more recent quantum theory calls a halt to this motion as soon as the nucleus and electron have approached to within a distance of 0.53×10^{-8} centimetres of one another, and by so doing keeps the universe in being as a going concern (p. 127). Other halts are also established at 4, 9, 16, etc. times this distance, but here the prohibition on further progress is not absolute. At these longer distances the demand of the quantum theory "thus far shalt thou go and no further," seems to be replaced by "thou shalt go no further until after a long time." And it now seems possible, on the astronomical evidence, that the prohibition at the shorter distance may not be absolute either. From the physical end nothing is known for certain, although here again it seems contrary to the newer conception of physics, as embodied in the wave-mechanics, that any such absolute prohibition should exist, either for the hydrogen atom or for other more complex atoms. Perhaps after waiting for a long time in the orbit nearest to the nucleus, the electron is permitted, or even encouraged or compelled, to proceed; it merges itself into the nucleus and a flash of radiation is born in a star. This provides the most obvious mechanism for the annihilation of electrons and protons which the evidence of astronomy seems to demand. It will, however, be clearly understood that this is a purely conjectural conception of the mechanism; we shall return to a further consideration of this very intricate problem in Chapter V.

If this conjecture should prove to be sound, not only the atoms which provide stellar light and heat, but also every atom in the universe, are doomed to destruction, and must in time dissolve away in radiation. The solid earth and the eternal hills will melt away as surely, although not as rapidly, as the stars:

> The cloud-capped towers, the gorgeous palaces,
> The solemn temples, the great globe itself,
> Yea, all which it inherit, shall dissolve,
> And . . . leave not a rack behind.

And if the universe amounts to nothing more than this, shall we carry on the quotation:

> We are such stuff
> As dreams are made on; and our little life
> Is rounded with a sleep,

—or shall we not?

CHAPTER IV

Carving out the Universe

We have commented on the surprising emptiness of space: six specks of dust in Waterloo Station about represent the extent to which it is occupied by stars in its most crowded parts. The comment might well have taken another form. Six specks of dust contain, let us say, a thousand million million molecules. Our model of space is empty because this great number of molecules happens all to be aggregated into as few as six lumps. In real space the unit of aggregation is the star, and an average star contains about 10^{56} molecules—a number so large that it is quite useless to try to imagine it. The emptiness of space does not originate from any paucity of molecules; it originates from the circumstance that, apart from those which form the tenuous clouds of gas stretching from star to star, the molecules are aggregated together in the huge colonies we call stars, with about 10^{56} members to each. Why should the molecules in space herd together in this way, when the molecules in the rooms in which I am writing and you are reading do not?

Following a well-tried scientific method, we may attempt to discover why these aggregates have formed, by first examining what keeps them together now that they have formed. The earth's atmosphere consists of about 10^{41} molecules. Why do they stay pressed down into an atmosphere instead of spreading out through space? The answer is of course provided by the earth's gravitation. A bullet fired from the earth's surface with a speed of 6.93 miles a second or more will fly off into space, because the earth's gravitational pull is inadequate to hold it back when it moves with so high a speed. But a bullet fired with a speed of less than 6.93 miles a second does not leave the earth; its speed is inadequate to take it clear of the earth's pull. Thus the molecule-bullets which form the earth's atmosphere, flying with speeds less than a third of a mile a second, have no chance at all of getting away. The earth's gravitation continually pulls them back to earth, so that the earth retains its covering of air.

At rare intervals a molecule may experience a succession of exceptionally lucky collisions with other molecules, and so attain a speed of more than 6.93 miles a second. A molecule which arrives at the outside of the earth's atmosphere with such a speed will leave the earth altogether, and join the interstellar crowd of stray molecules. The earth is continually shedding its atmosphere in this way, but calculation shews that the loss, even in millions of millions of years, is quite insignificant, so that we may regard the earth's atmosphere as permanent.

It is the same with the sun. The sun's heat has broken up the molecules of its atmosphere into their constituent atoms, and these move with an average speed of about 2 miles a second. But an atom-bullet would have to move at about 380 miles a second to escape altogether from the sun, so that the solar atoms remain to form an atmosphere.

If all the molecules of air in an ordinary room were collected into a bunch at the centre of the room, the ball of air so formed would of course exert a gravitational pull on its outermost molecules, of the same kind as the earth and sun exert on the molecules of their atmospheres. But, because the weight of this ball of air is so small, the intensity of its gravitational pull would also be small; indeed it would be so feeble that a speed of about a yard a century would be enough to take the outermost molecules clear of it. As the molecules of ordinary air move at about 500 yards a second, such a ball of air would immediately scatter through the whole room. On the other hand, if the room were big enough to contain the sun, all its molecules could stay in a ball at the centre, just as they do in the sun. The outermost molecules would need a speed of at least 380 miles a second to escape, so that their actual speeds of 500 yards a second or so would be of no service to them.

Planetary Atmospheres. In general the question of escape or no escape depends on the outcome of a battle between the molecular speeds of the outermost molecules, and the intensity of the gravitational hold which the remainder of the mass exerts on them. The solar system provides many examples of this. The moon has only a sixth as much gravitational hold over the molecules of an atmosphere as the earth has, with the result that any atmosphere the moon may ever have had, has escaped by now. Mercury has two-fifths of the earth's gravitational hold, but, owing to its nearness to the sun, its sunward surface is very hot, with the consequence that its atmosphere also has escaped. The gravitational hold of Mars on its molecules, is only a fifth of the earth's, but its surface is cooler. Calculation shews that water-vapour and heavier molecules ought to remain, while the lighter molecules of helium and hydrogen ought to have escaped. This probably represents what has actually happened. The largest satellite of Saturn and the two largest satellites of Jupiter would exercise about the same gravitational hold as the moon, but as their surfaces must be enormously colder than that of the moon, they ought to be able to retain atmospheres. Some observers claim to have seen indications of atmospheres on all three satellites. All the four major planets exert stronger gravitational holds over their molecules than the earth, and so retain their atmospheres with ease, while Venus, with approximately the same gravitational hold as the earth, also retains an atmosphere.

These considerations amply explain why the molecules of the stars must necessarily remain aggregated now that the aggregates have once been formed, but the question of how and why these aggregates formed in the first instance is far more complex. What, for instance, determined that there should be about 10^{56} molecules in each star rather than 10^{54} or 10^{58}?

Gravitational Instability

It is natural to enquire whether the forces which now keep a star together may not have been also responsible for its falling together in the first instance. This leads us to study the aggregating power of gravitation in some detail.

Five years after Newton had published his law of gravitation, Bentley, the Master of Trinity College, wrote him, raising the question of whether the newly discovered force of gravitation would not account for the aggregation of matter into stars, and we find Newton replying, in a letter of date December 10, 1692:

It seems to me, that if the matter of our sun and planets, and all the matter of the universe, were evenly scattered throughout all the heavens, and every particle had an innate gravity towards all the rest, and the whole space throughout which this matter was scattered, was finite, the matter on the outside of this space would by its gravity tend towards all the matter on the inside, and by consequence fall down into the middle of the whole space, and there compose one great spherical mass. But if the matter were evenly disposed throughout an infinite space, it could never convene into one mass; but some of it would convene into one mass and some into another, so as to make an infinite number of great masses, scattered great distances from one to another throughout all that infinite space. And thus might the sun and fixed stars be formed, supposing the matter were of a lucid nature.

An exact mathematical investigation on which I embarked in 1901, not only confirms Newton's conjecture in general terms, but also provides a method for calculating what size of aggregates would be formed under the action of gravitation.

The Formation of Condensations. You stand in the middle of a room and clap your hands. In common language you are making a noise; the physicist, in his professional capacity, would say you are creating waves of sound. As they approach one another, your hands expel the intervening molecules of air. These stampede out, colliding with the molecules of outer layers of air, which are in turn driven away to collide with still more remote layers; the disturbance originally created by the motion of your hands is carried on in the form of a wave. Although the individual molecules have an average speed of 500 yards a second, the zig-zag quality of their motion reduces the speed of the disturbance, as we have already seen, to about 370 yards a second—the ordinary velocity of sound. As the disturbance reaches any point the number of molecules there becomes abnormally high, for the stampeding molecules add to the normal quota of molecules at the point. This of course produces an excess of pressure. It is this excess pressure acting on my ear-drum that transmits a sensation to my brain, so that I hear the noise of your clapping your hands.

This excess of pressure cannot of course persist for long, so that the excess of molecules which produces it must rapidly dissipate. It is thus that the wave passes on. Yet there is one factor which militates against its dissipation. Each molecule exerts a gravitational pull on all its neighbours, so that where there is an excess of molecules, there is also an excess of gravitational force. In an ordinary sound wave this is of absolutely inappreciable amount, yet such as it is, it provides a tiny force holding the molecules back, and preventing them scattering as freely as they otherwise would do. When the same phenomenon occurs on the astronomical scale, the corresponding forces may become of overwhelming importance.

Let us speak of the gas in any region of space where the number of molecules is above the average of the surrounding space, as a "condensation." Then it can be proved that, if a condensation is of sufficient extent, the excess of gravitational force may be sufficient to inhibit scattering altogether. In such a case, the condensation may continually grow through attracting molecules into it from outside, whose molecular speeds are then inadequate to carry them away again.

Whether this happens or not will depend of course on the speed of molecular motion in the gas, as well as on the size of the condensation. But it will not depend at all on the extent to which the process of condensation has proceeded. By doubling the excess number of molecules in any condensation, we double the extent to which condensation has proceeded. In so doing, we double the gravitational pull tending to increase the condensation, but we also double the excess pressure which tends to dissipate it; we double the weights on each side of the balance, but the balance still swings in the same direction. If once conditions are favourable to its growth, a condensation goes on growing automatically until there are no further molecules left for it to absorb.

The greater the extent in space of a condensation, the more favourable conditions are to its continued growth. Other things being equal, a condensation two million miles in diameter will exert twice the gravitational force of a condensation one million miles in diameter, but the excess pressures are the same in the two cases. Thus, the larger a condensation is the more likely it is to go on growing, and by passing in imagination to larger and larger condensations we must in time come to condensations of such a size that they are bound to keep on growing. Nature's law here is one of unrestricted competition. Nothing succeeds like success, and so we find that condensations which are big to start with have the capacity of increasing still further, while those which are small merely dissipate away.

Suppose now that an enormous mass of uniform gas extends through space for millions of millions of miles in every direction. Any disturbance which destroys its uniformity may be regarded as setting up condensations of every conceivable size.

This may not seem obvious at first; it may be thought that a disturbance which only affected a small area of gas would only produce a condensation of small extent. Such an argument overlooks the way in which the gravitational pull of a small body acts throughout the universe. The moon raises tides on the distant earth, and also tides, although incomparably less in amount, on the most distant of stars. Each time the child throws its toy out of its baby-carriage, it disturbs the motion of every star in the universe. So long as gravitation acts, no disturbance can be confined to any area less than the whole of space. The more violent the disturbance which creates them, the more intense the condensations will be to begin with, but even the smallest disturbance must set up condensations, although these may be of extremely feeble intensity. And we have seen that the fate of a condensation is not determined by its intensity but by its size. No matter how feeble their original intensity may have been, the big condensations go on growing, the small ones disappear. In time nothing is left but a collection of big condensations. The mathematical analysis already referred to shews that there is a definite minimum weight such that all condensations below this weight merely dissipate away into space. To a good enough approximation for our present purpose, this minimum weight is such that if a tent of this weight of gas were isolated in space, and all the rest of the gas annihilated, the molecules would just and only just fail to escape from its surface.*

We may say that the original uniformly distributed mass of gas was "unstable" because any disturbance, however slight, causes it to change its configuration entirely; it had the dynamical attributes of a stick balanced on its point, or of a soap-bubble which is just ready to burst.

* This is near enough, but not absolutely accurate. Exact mathematical analysis shews that the weight of the minimum condensation M is given by

$$M = (\tfrac{1}{3}\pi\kappa)^{\tfrac{3}{2}} \frac{C^3}{\gamma^{\tfrac{3}{2}}\rho^{\tfrac{1}{2}}},$$

where C, γ, ρ, κ are the molecular velocity, gravitation constant, initial density, and ratio of specific heats, whereas the weight from which molecules moving with velocity C just fail to escape is given by

$$M = \frac{3}{4\pi} \frac{C^3}{\gamma^{\tfrac{3}{2}}\rho^{\tfrac{1}{2}}}.$$

With $\kappa = 1\tfrac{2}{3}$ the minimum weight of condensation is 9.7 times the weight which is just adequate to retain the molecules.

Primaeval Chaos. These general theoretical results may now be applied to any mass of gas we please. Let us begin by applying them to Newton's hypothetical "matter evenly disposed throughout an infinite space." We return in imagination to a time when all the substance of the present stars and nebulae was spread uniformly throughout space; in brief, we start from the primaeval chaos from which most scientific theories of cosmogony have started. Hubble has estimated that if the whole of the matter in those parts of the universe we know, were redistributed evenly throughout space, the gas so formed would have only about 1.5×10^{-31} times the density of water. This estimate is almost certainly on the low side, even as representing present conditions, and in trying to reconstruct the primaeval gas we must add something to allow for the molecules and atoms which have melted away into radiation in the intervening period. On the whole, perhaps 10^{-30} is not an unreasonable density to assign to the hypothetical primaeval nebula. It is almost inconceivably low. In ordinary air, at a density of one eight-hundredth that of water, the average distance between adjoining molecules is about an eight-millionth part of an inch; in the primaeval gas we are now considering, the corresponding distance is two or three yards. The contrast again leads back to the theme of the extreme emptiness of space.

What is the minimum weight of condensation that would persist in this primaeval gas?

Calculation shews that if ordinary air were attenuated to this extraordinary degree, no condensation could persist and continue to grow unless it had at least $62\frac{1}{2}$ millon times the weight of the sun; any smaller weight of gas would exert so slight a gravitational pull on its outermost molecules, that their normal molecular speeds of 500 yards a second

CARVING OUT THE UNIVERSE

would lead to the prompt dissipation of the whole condensation.

We can carry out similar calculations with reference to other assumed densities of gas, and other molecular velocities. The following table shews the weights of condensations which would be formed in primaeval masses of chaotic gas having the densities shewn in the first column, and the various molecular velocities mentioned at the heads of the remaining columns. In each case the weights of the condensations are given in terms of the weight of the sun:

Density in terms of water	Mol. vel. of 500 yards a sec.	Mol. vel. of 1000 yards a sec.	Mol. vel. of 2000 yards a sec.	Mol. vel. of 3000 yards a sec.
10^{-29}	25,000,000	200,000,000	1,500,000,000	5,000,000,000
10^{-30}	62,500,000	500,000,000	4,000,000,000	13,000,000,000
1.5×10^{-31}	160,000,000	1,300,000,000	10,000,000,000	30,000,000,000

All known stars have weights comparable with that of the sun. Thus if, as Newton conjectured, the stars first came into being as condensations of this kind, then the entries in this table ought to be comparable with unity. Newton's conjecture, in the form in which we have just considered it, is clearly untenable, since all the calculated weights are many millions of times that of the sun. If there ever existed a primaeval chaos of the kind we are now considering, it would not condense into stars, but into enormously more massive condensations, each having the weight of millions of stars.

The Birth of the Great Nebulae

Now it is significant that bodies are known in space having weights equal to those just calculated, namely the great extra-galactic nebulae. There are two nebulae whose weights can be determined with fair accuracy, namely the Great Nebula in Andromeda (Plate IV, p. 30) and the nebula N.G.C. 4594 in Virgo (Plate XV). Hubble estimates these to be as follows:

Nebula M. 31:   weight = 3500 million times that of sun
" N.G.C. 4594:   "   = 2000   "   "   "

These estimates are again probably both on the low side, but their general order of magnitude is such as to suggest that the condensations which would first be formed out of the primaeval nebula must have been the great extra-galactic nebulae, and not mere stars. It is of course at best only a conjecture that the great nebulae were formed in this manner —if for no other reason because we can never know whether the hypothetical primaeval nebula even existed—but it seems the most reasonable hypothesis we can frame to explain the fact that the present nebulae exist. These nebulae are so generally similar to one another that it seems likely that they must all have been produced by the action of the same agency, and that which we have just considered provides a reasonable explanation which, apart from the postulated existence of the continuous primaeval nebula, is based on *verae causae*.

The great nebulae are of course not exactly similar, and our next inquiry must be as to the origin of their differences.

If the condensations in the primaeval gaseous nebula had formed and contracted in an absolutely regular fashion, the final product would be an array of perfectly equal and similar masses of gas spaced with perfect regularity. But nature is seldom as regular as this; and we need not be surprised that the observed nebular array is not evenly spaced, or that its members are neither equal in weight, nor symmetrically arranged. As the original condensations in the

PLATE XV

The Nebula N.G.C. 4594 in Virgo.

The Nebula N.G.C. 7217.

Mt. Wilson Observatory.

primaeval gas contracted, they must have produced currents, and these would hardly be likely to occur absolutely symmetrically. If the motion in each mass of condensing gas had been directly towards the centre of the condensation at every point, the final result would have been a spherical nebula devoid of all motion, but any less symmetrical system of currents would result in a spin being given to each contracting mass. This spin would no doubt be very slow at first, but the well-known principle of "conservation of angular momentum" requires that, as a spinning body contracts, its rate of spin must increase. Thus when the process of condensation was complete, the final product would be a series of nebulae rotating at different rates.

Nebular Rotation. And this is exactly what is observed; so far as our evidence goes the nebulae are in rotation, and at different rates. The various parts of the surface of any rotating mass necessarily have different speeds in space. The sun for instance rotates about its axis in such a direction that the surface we see is moving always from east to west; as a result the eastern limb is always advancing towards the earth, while the western limb is receding from us. A spectroscope turned onto different parts of the sun's surface in succession at once reveals these differences of speed; they not only assure us of the sun's rotation, but enable us to measure its amount. The nebulae may be examined in the same way, and the examination shews that a large number of them are rotating with the perfectly regular motion of a solid body—a spinning top, for instance. Measured by terrestrial standards their rates of rotation seem extraordinarily slow; for instance the Great Nebula (M 31) in Andromeda requires about 19,000,000 years to make a complete rotation, but this apparent slowness is an inevitable result of the huge size of the nebula. Even to get round once in 19,000,000 years, the outer parts of the nebula have to move with speeds of hundreds of miles a second.

A few of the nebulae are quite irregular in shape, but the majority have regular shapes, and it is highly significant that these are precisely the shapes which, it can be calculated mathematically, would be exhibited by rotating masses of gas. Actually there is a far stronger case than this for supposing the nebulae to be rotating masses of gas. From the purely observational evidence of surface-brightness and other characteristics, Hubble found that nearly all of these nebulae could be arranged in a single linear sequence—they could be arranged in order like beads on a string. And this order proved to be practically identical with the sequence which had previously been calculated, by purely theoretical methods, for the configurations of masses of gas rotating at gradually increasing rates of speed.

Let us examine this sequence of theoretical configurations in their natural order.

A mass of gas which was not rotating at all would of course assume a spherical shape under its own gravitation. A number of perfectly spherical nebulae are known; a typical example is shewn in fig. 1 on Plate XVI.

With slight rotation the mass assumes the shape of a slightly flattened orange, like the earth or Jupiter. Nebulae of this shape are also known in abundance; an example is shewn in fig. 2 on the same plate.

With a higher degree of rotation the degree of flattening increases, but theoretical calculation shews that the orange shape is soon departed from. The equator first begins to shew a pronounced bulge, until finally, with sufficient rota-

PLATE XVI

Fig. 1.
N.G.C. 3379.

Fig. 2.
N.G.C. 4621.

Fig. 3.
N.G.C. 3115.

Fig. 4.
N.G.C. 4594.

Fig. 5.
N.G.C. 4565.

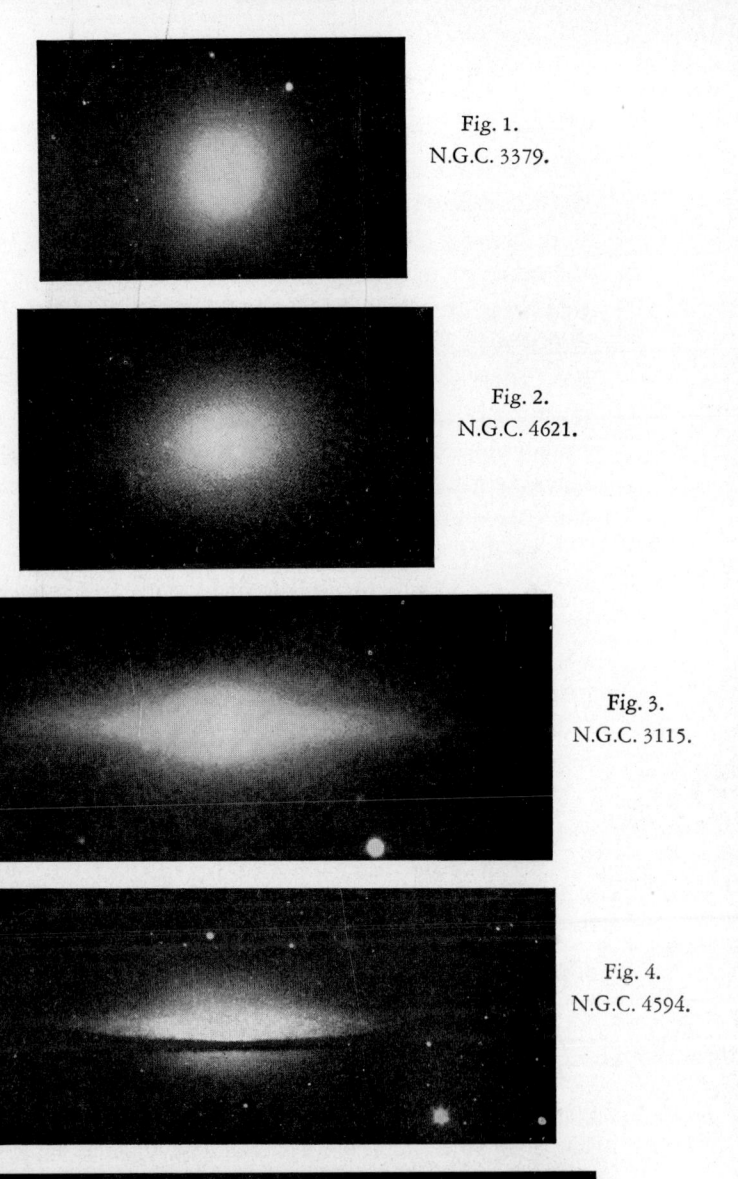

Mt. Wilson Observatory.

A sequence of Nebular Configurations.

tion, this develops into a sharp edge, the rotating mass now being shaped like a double-convex lens. This prediction of theory is abundantly confirmed by observation, a large number of these lens-shaped nebulae being observed in the sky. An example is shewn in fig. 3 on Plate XVI.

The next step is somewhat sensational. Further rotation does not, as might be expected, result in still further flattening. Up to now, each increase in rotation has made the bulge on the equator sharper, but this is now as sharp as it can be. Theory shews that the flattening has also proceeded to the utmost possible limit, and that the next stage must consist in matter being ejected through the sharp edge of the equator and spread throughout the equatorial plane. Here again observation confirms theory; figs. 4 and 5 (Plate XVI) shew types of nebulae actually observed, the former being the nebula in Virgo which we have already had under discussion.

The comparatively thin layer of gas which now lies in the equatorial plane is similar in one respect at least to Newton's matter "evenly disposed throughout an infinite space." Disturbances can be set up in it in a variety of ways, and any disturbance, no matter how slight, must result in the creation of a series of condensations. As before, those below a certain limit of size disappear of themselves, while those above this limit continually increase in intensity until they have all the gas in the equatorial plane absorbed. Again, as with the hypothetical primaeval chaos, we can calculate the minimum size of condensation which can be expected to have a permanent existence, and once again the result proves to be highly significant.

Hubble's estimates of the total weights of two conspicuous

nebulae have already been given. As the distances, and therefore also the sizes, of both these nebulae are known, it is an easy matter to calculate the average density of the gas throughout the whole nebula. The average density in M 31 is found to be about 5×10^{-22} of that of water; the corresponding number for N.G.C. 4594 is 2×10^{-22}. These figures give us some idea of the density of matter in the outer regions of the nebula. Although these densities are about a thousand million times as great as the estimated density of the original primaeval nebula of space, they are still almost inconceivably low. There is still only about one molecule to the cubic inch, and a single breath from the lungs of a fly could fill a large cathedral with air of this density.

On proceeding to calculate the weights of the smallest condensations which could form and persist in a gas of this low density, we obtain the results shewn in the following table. The molecular velocities are taken rather low, so as to allow for the cooling which must occur when the gas is spread out in the equatorial plane of the nebula.

Again the weights of the condensations are given in terms of the weight of the sun. And the significant fact emerges that most of the entries in the table represent weights comparable with that of the sun. We are dealing with stellar weights at last; the condensations which must form in the outer regions of the great nebulae will have weights comparable with those of the stars.

Density in terms of water	Mol. vel. of 100 yards a sec.	Mol. vel. of 300 yards a sec.	Mol. vel. of 500 yards a sec.
10^{-21}	1.7	36	220
10^{-22}	5	130	625
10^{-23}	17	360	2200

PLATE XVII

Mt. Wilson Observatory.

The "Whirlpool Nebula" (*M* 51) in Canes Venatici.

PLATE XVIII

Mt. Wilson Observatory.

The Nebula *M* 81 in Ursa Major.

CARVING OUT THE UNIVERSE

The Birth of Stars

And indeed there can be but little doubt that the process we have just been considering is that of the birth of stars. Even a casual glance at photographs of nebulae suffices to shew that the matter which has been ejected into the equatorial plane of a nebula does not lie uniformly spread out in that plane; it is seen to have fallen into bunches, knots or condensations. These are apparent enough in many of the nebular photographs already shewn, but they can be seen still more clearly in nebulae which are viewed nearly full on, such as for instance the two striking nebulae shewn in Plates XVII and XVIII.

These bunches are invariably too large to be interpreted as single stars; they are more probably groups of stars. In the largest telescopes they break up into great numbers of points of light in the way already exhibited in Plate XI (p. 66). We have already mentioned the reasons which compel us to regard these points of light as actual stars, the principal being that some of them shew the characteristic light-fluctuations of the Cepheid variables. It is not altogether clear whether the stars are formed directly as condensations in the equatorial plane of the nebula, or whether larger condensations form first, namely the bunches observable in nebular photographs, which subsequently form smaller condensations, the stars. On the whole it seems likely that there are two processes involved—first the break-up of the nebular matter into big condensations, and then the break-up of these big condensations into stars. Such a succession of processes might well accompany a gradual cooling of the matter, and it is of course possible that there are even more than two processes involved. There is no need to form a final opinion on this at present, as it is in no way essential to the progress of the main argument.

A collection of nebular photographs enables us to follow nebular evolution from the earliest stages shewn in Plate XVI (p. 196), through the first appearance of granular bunches, such as are shewn in Plate XVII, and the first distinct appearance of stars shewn in Plate XVIII, down to the later stages, such as are shewn in Plates XIX and XX, in which the nebula appears to be but little more than a cloud of stars. Hubble has found it possible to follow the sequence still further, and can trace a continuous transition from the nebulae of this last type to pure star-clouds such as the Greater and Lesser Magellanic Clouds shewn in Plate XXI.

Thus the stars appear to have been born in much the same way as we have conjectured that their parents, the great nebulae, had been born before them, namely, through the agency of what is generally known as "Gravitational Instability." This causes any mass of chaotic gas to break up into detached condensations, and, the more tenuous the original gas, the greater the weights of the condensations formed out of it. The original primaeval nebula was of such low density that the condensations which formed in it weighed thousands of millions of times as much as the sun. These increased their density so much in contracting that when their rotation caused them to eject gaseous matter, this condensed into masses of stellar weight which we believe actually to be stars.

We have less certain knowledge of the former process than of the latter. Our only reason for thinking that the former process ever occurred is that the extra-galactic nebulae now exist. There is no evidence that the primaeval chaotic

PLATE XIX

Mt. Wilson Observatory.

The Nebula *M* 101 in Ursa Major.

PLATE XX

Mt. Wilson Observatory.

The Nebula *M* 33 in Triangulum.

nebula ever existed, beyond the fact that the hypothesis of its previous existence leads to a very satisfactory explanation of the present nebulae existing as they now do. On the other hand, we not only know that the stars exist: we also know that the masses of gas exist out of which theory shews that stars must necessarily be born. They are the tenuous equatorial fringes of the great nebulae. Our telescopes shew us both the nebular fringes and the stars, and we can almost study the actual process of birth.

The Galactic System of Stars. If this is the true account of the birth of the stars, then our sun and its companions in space must have been born out of a rotating nebula. Observation gives strong support to this conclusion. Since the time of the Herschels, it has been a matter of frequent comment that the galactic system has the general shape of the extra-galactic nebulae, the galactic plane of course representing the equatorial plane of the original nebula. On purely observational grounds, present-day astronomical thought is moving rapidly towards regarding the whole galactic system either as a rotating nebula or the remains of one. It is even possible that this may still retain a central region which is as yet uncondensed into stars. In the direction of the constellations Scorpio and Ophiuchus are dark clouds which may either veil the centre of the system or may conceivably be the centre itself.

In 1904 Kapteyn found that the directions of motion of the stars in the vicinity of the sun were not distributed at random. The stars appeared to prefer to move to and fro along a certain direction in the galactic plane rather than in other directions—"star-streaming," he called it. This peculiarity in the motion of the stars may be expected to throw some light on their origin.

Each star moves in a complicated orbit under the gravitational attraction of all the other stars of the galactic system. It is not possible to calculate this orbit in detail. The orbit of a planet round the sun is easily calculated because only two bodies are involved, the planet and the sun. But even when there are only three bodies involved, it is impossible to calculate the orbits that each describes under the attractions of the other two jointly: this is the famous problem of three bodies, which has never been solved. When, as in the galactic system, thousands of millions of stars are involved, it is naturally useless to try to calculate the orbit of each star —it would be as futile as trying to calculate the path of each molecule in a gas.

Yet the same statistical methods which give us useful information as to the properties of a gas may be applied to studying the motions of the stars. There are so many stars that we do not trouble about individuals at all, we just treat them all together as a crowd. To treat them as individuals would be as though the railway company tried to forecast the Bank Holiday traffic between London and Brighton by considering the finances, habits and psychology of each individual Londoner.

Without going into individual details, we can see that each star must describe an orbit which, after touring round a large part of the galaxy, comes back to somewhere near its starting point. Calculation shews that each such circuit must take hundreds of millions of years to complete. Even so, the stars will mostly have performed several complete circuits while the earth has been in existence, and if we are right in supposing the ages of the stars to be millions of millions of years, each star must have toured round the galaxy several thousands of times. We should accordingly expect the galaxy to have assumed a definite permanent shape by now; the distribution of stars in its different parts ought to have become something like steady, and the stars ought to have settled down to a state approximating to one of steady motion.

Statistical methods of investigation shew that there is not a great number of possible arrangements for a system of stars which has lived long enough to attain a steady state. If the system as a whole has no rotation at all, there is only one arrangement; the stars form a globular mass with perfect symmetry in all directions. The observed globular clusters (Plate IX) provide good approximations to this type of formation, although Shapley has found that the majority are not absolutely spherical in shape. If the system as a whole is endowed with rotation, the possible configurations are all of a flattened and symmetrical shape, like a coin, a watch or a round biscuit—in other words a system of stars in rotation must be shaped pretty much as we believe the galaxy to be shaped. Furthermore the motions of these stars must shew "star-streaming" of precisely the kind discovered by Kapteyn.

Thus both the shape of the galaxy and the peculiarities of motion of its stars indicate that the galactic system as a whole must be in a state of rotation. Recent observational researches by Oort, Plaskett and others make it fairly certain that the rotation required by theory is an actual fact. The motions of the stars indicate that the whole galactic system is rotating at the rate of about one revolution every 300 million years. And the hub of this gigantic wheel is found to coincide very closely, both in its direction and distance, with the spot which Shapley had previously fixed as the geometrical centre of the galactic system from his researches on the distribution of the globular clusters.

THE UNIVERSE AROUND US

Thus, since rotation cannot be generated out of nothing, all the phenomena agree in shewing that the galactic system must have been born out of a rotating body. We are acquainted with only one type of astronomical body which is of sufficient size to turn into a galactic system, namely the great nebulae, and as the majority of these are believed, and some are known with certainty, to be in rotation, it seems reasonable to conclude that the galactic system must have been born out of a nebula, unless indeed its structure is still such that we should even now describe it as a nebula if we saw it from the great distance from which we view the other great nebulae. The observed period of rotation of the galactic system, about 300 million years, is substantially longer than the period, either known or suspected, of any of the nebulae, but the dimensions of the galactic system are also greater than those of any known nebula, and the two facts hang together. Again, the number of stars in the galactic system is probably substantially higher than in any nebula, as also is the total weight of these stars.* All this makes it clear that if the galaxy is, or ever has been, one of the great nebulae, it must have been one of unusual size and weight.

We have seen how the sun and all the stars are continually losing weight as the result of their emission of radiation. It follows that the total weight of the galactic system is for ever decreasing, and as a consequence its gravitational hold on its constituent stars is continually weakening.

---

* The following estimates have already been mentioned:

Number of stars in Galaxy (Seares)	30,000,000,000
" " (Shapley)	100,000,000,000
Weight of Galaxy in terms of sun (Eddington)	270,000,000,000
" nebula M 31 in terms of sun (Hubble)	3,500,000,000
" " N.G.C. 4594 in terms of sun (Hubble)	2,000,000,000

PLATE XXI

Harvard (Arequipa) Observatory.
The Lesser Magellanic Cloud.

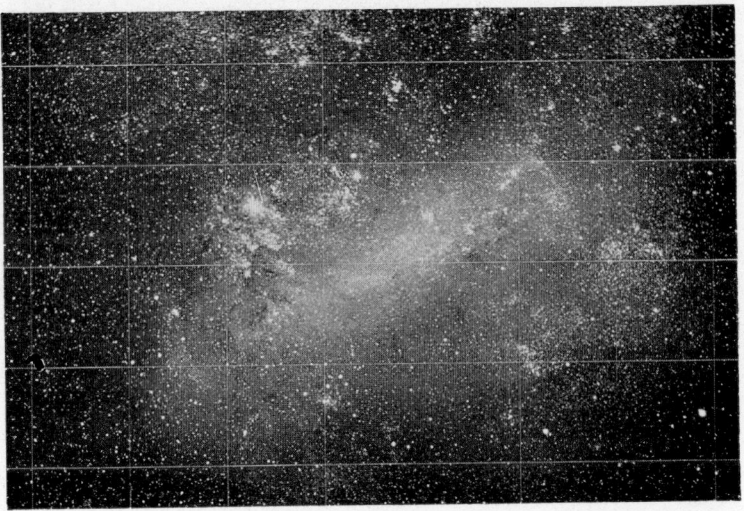

Franklin-Adams Chart.
The Greater Magellanic Cloud.

CARVING OUT THE UNIVERSE

If this gravitational hold were suddenly to vanish altogether, each star would replace its present curved path by a perfectly straight line, along which it would travel at its present speed, undeflected by any gravitational forces from other stars, so that the stars which now constitute the galactic system would soon be scattered through the whole of space. In brief, if the gravitational pull of the stars were suddenly abolished, the galaxy would begin to expand at a great rate.

Although this is not likely to happen, the gradual abolition of the gravitational pull of the stars, as they turn their weight into radiation, must cause the galaxy to expand all the time at a slow rate: calculation suggests that its present rate of expansion would double its size in about 30 million million years. The expansion must have been far more rapid in the past, when the stars were full of youthful vigour and squandered their substance more lavishly than now, so that it seems probable that the galactic system was substantially smaller and more compact in the past than now, and the original nebula probably smaller still.

We have seen how the stars in the great nebulae appear to be congregated in bunches or clusters. The globular clusters in the galactic system may possibly be bunches of stars of the same general type, which have remained undisturbed by other groups of stars and so have assumed the globular form under their own attraction—just as a mass of gas would do. Shapley finds that these clusters lie somewhat outside the galactic plane; it looks as though they were broken up or disorganised in travelling through this plane, where they would encounter other stars.

By contrast groups of stars of the type generally described as moving clusters—the Pleiades, the Hyades, the stars of the Great Bear and a crowd of others voyaging in company with them through space—are generally found to move in the galactic plane. These may quite possibly represent the final vestiges of globular clusters which have been broken up by interaction with other stars, all except the most massive members having been knocked out of formation. Mathematical analysis shews that the interaction between the stars of such moving clusters and other stars in the galactic plane would cause each cluster to assume the shape of a flat biscuit or watch, of diameter equal to 2½ times its thickness. It is significant that the majority of the moving clusters shew a flattening of this kind, its amount agreeing tolerably well with the calculated value. It is even conceivable that the "local cluster" surrounding the sun (p. 62) may be the remains of such a bunch of stars.

The motions of these clusters may also induce a further flattening, in a direction perpendicular to their motion. Some clusters shew this further flattening, the Ursa Major cluster being a striking example.

The Birth of Binary Systems

In discussing the way in which nebulae might be born out of chaos, we noticed that the existence of currents in the primordial medium would endow the resulting nebulae with varying amounts of rotation. For the same reason the children of the nebulae, the stars, must also be endowed with rotation at their birth. There is a further reason for such rotation. The general principle of the "conservation of angular momentum" requires that rotation, like energy, cannot entirely disappear. Its total amount is conserved, so that when a nebula breaks up into stars, the original rotation of the nebula must be conserved in the rotations of the stars. Thus the stars, as soon as they come into being, are endowed with rotations transmitted to them by their parent nebula, in addition to the rotations resulting from the currents set up in the process of condensation.

Their continual loss of weight causes the physical conditions of the stars to change, and we shall find in the next chapter that this change generally involves a shrinkage of the star's diameter. The same principle of "conservation of angular momentum" now requires that, as a star shrinks, its speed of rotation shall increase. In brief, as a star ages, it spins faster and faster.

Now rotation was the essential factor in the birth of the stars out of the parent nebula. A nebula perfectly devoid of rotation would not, so far as we can see, break up into stars at all, and this prediction of theory appears to be confirmed by observation, since nebulae of the perfectly spherical type shewn in fig. 1 of Plate XVI can never be resolved into stars in the telescope. On the other hand we saw how nebulae which were initially endowed with rotation would continually increase their speed of rotation under shrinkage, until finally their rotation broke them up and produced a family of stars out of each. The question now obviously arises whether, as the speed of rotation of the stars increases, these are likely to break up in their turn, and produce yet a third generation of astronomical bodies. Again we might expect that mathematical analysis would apply to large and small bodies equally, irrespective of scale. And a detailed examination of the problem shews that in actual fact the process we have had under consideration would repeat itself, and again bring a further generation of smaller bodies into being, provided the physical conditions were suitable.

The physical conditions, however, prove not to be suitable; they certainly fail in one respect at least. Although a rotating star may eject gaseous matter in its equatorial plane, the whole process will be on a much smaller scale than in the nebulae. We might expect the ejected matter to form condensations as before, but calculation shews that, unless the molecular velocity is extraordinarily low, no condensation can survive unless it has a weight greater than the whole weight of the star! This means that with any reasonable molecular velocity, the ejected gas would not form condensations at all. It would merely scatter into the surrounding space, forming an atmosphere without any distinct condensations.

Such is the course of events if the stars, like the nebulae before them, are treated as pure masses of gas. Another alternative must, however, be considered.

The Fission of Liquid Stars. We have seen how a gaseous nebula devoid of rotation would assume a strictly spherical shape under its own gravitational attraction, while slight rotation would cause it to flatten into an orange shape, like the earth. The earth also has assumed this shape on account of its rotation, although its internal structure is very different from that of a gaseous nebula.

Strict mathematical investigation shews that this flattened-orange shape must be common to all slowly rotating bodies, regardless of their internal composition; gases, liquids, and plastic bodies assume it equally. But the shape of a rapidly rotating body must depend very greatly on its internal arrangement and constitution, being especially affected by the extent to which the weight of the body is concentrated near its centre.

As a consequence of the high compressibility of gases, this central concentration of weight reaches its extreme limit in a purely gaseous mass. The opposite extreme is reached in a mass of uniform incompressible liquid such as water, in which there can be no central concentration at all. As a mass of this latter type increases its speed of rotation, the slightly flattened-orange shape merely gives place to the shape of a more flattened orange. The tendency of a gaseous mass to form a sharp edge round the equator is entirely absent, and the cross-section of its figure remains elliptical throughout. At a still higher speed of rotation, the equator loses its circular shape and it too becomes elliptical. The figure has now three unequal diameters, but every cross-section is strictly elliptical; the figure is an "ellipsoid." After this, its longest diameter begins to elongate until the mass, still ellipsoidal in shape, has formed a cigar-shaped figure with a length nearly three times its shortest diameter.

A new series of events now begins. The mass of liquid gradually concentrates about two distinct points on its longest diameter, a waist or furrow forming across its middle. This furrow gets deeper and deeper until it has cut the body into two distinct detached masses, which now rotate in orbital motion about one another and form a binary star. The sequence of events is shewn in fig. 11; diagrams of the final stage as represented by actual binary stars have already been given on p. 52.

For comparison the sequence of shapes assumed by a rotating mass of gas is shewn in fig. 12, this being identical with the sequence of observed nebular shapes which is actually observed, and is illustrated photographically in Plate XVI, p. 196.

The two chains of configurations shewn in figs. 11 and 12 represent, it will be remembered, the two extreme cases of a rotating body whose substance is distributed with complete uniformity, and of a rotating body whose substance is very highly condensed towards its centre. As the constitutions of actual astronomical bodies must lie somewhere between these two extremes, we might naturally expect such a body to follow a series of configurations intermediate between the

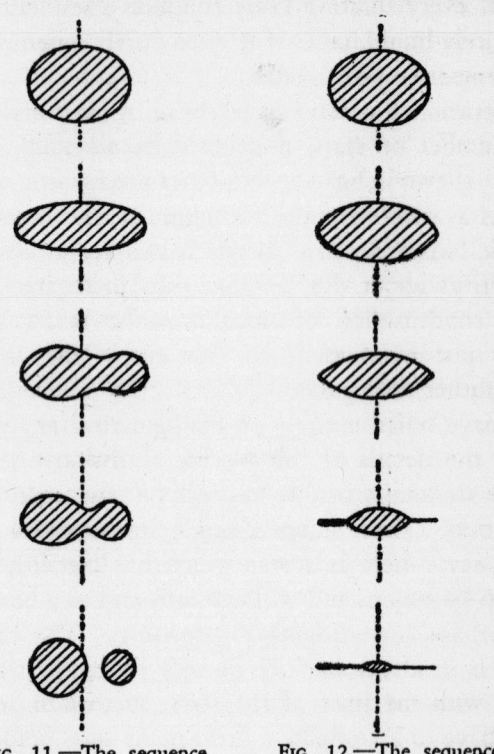

FIG. 11.—The sequence of configurations of a rotating mass of liquid.

FIG. 12.—The sequence of configurations of a rotating mass of gas.

two shewn in figs. 11 and 12. Theory shews that as a matter of fact it does not. All bodies having less than a certain critical degree of central condensation follow the sequence shewn in fig. 11, or a sequence differing only immaterially from this; all bodies having more than this critical amount of central condensation follow the sequence shewn in fig. 12. Thus when this critical degree of central condensation is reached there is a sudden swing over from fig. 11 to fig. 12. In brief, every rotating body conducts itself either as if it were purely liquid, or as if it were purely gaseous; there are no intermediate possibilities.

Observational astronomy leaves no room for doubt that a great number of stars, possibly even all stars, follow the sequence shewn in fig. 11. No other mechanism, so far as we know, is available for the formation of the numerous spectroscopic binary systems, in which two constituents describe small orbits about one another. In these stars, then, the central condensation of mass must be below the critical amount just mentioned; to this extent they behave like liquids rather than gases.

We have relied entirely on mathematical analysis in tracing out the details of the process of fission just described. And we are totally unable to check our theoretical results by observation. There is not a single star in the sky of which we can say:—here is a star which has certainly started to break up by fission, and will certainly end as a binary system. It is perhaps not altogether surprising. The breaking up process is in all probability of very short duration by comparison with the lives of the stars, so that in any case we should have to investigate a great many stars before catching one in the act of breaking into two.

On the other hand, a star in the act of breaking up ought to be very easily differentiated from ordinary stars. Mathematical analysis shews that its interior would be in a state of considerable turmoil, so that it would hardly be likely to shine with a steady light: it would be a "variable" star.

Further, its condition ought to shew a progressive change, although it is an open question whether this would be rapid enough to be detected in a few years of observation. Finally, if any group or class of stars were suspected of being stars in process of fission, it ought to be possible to arrange them in an order corresponding to the extent to which the fissional process had advanced, and the sequence so formed ought to end with stars in the physical condition of newly formed binaries.

I have recently suggested that the Cepheid variables, whose unknown mechanism of light variation renders such valuable service to the astronomer, are merely stars in the act of fission. Want of space prevents our entering here into the intricate question of how far they exhibit the peculiarities which mathematical analysis requires of stars in process of fission, but it is easily seen that they satisfy the three simple tests outlined above. They are certainly variable stars, and the light variations of different stars are so similar as to suggest very strongly that they all arise from the same cause. The periods of a number of Cepheids are suspected of change, and Hertzsprung has estimated that the prototype star, δ Cephei, which has now been observed for 126 years, is decreasing its period of light-fluctuation at the rate of about a tenth of a second per annum; thus a million years would reduce its present period of 5 1/3 days by over a day. Finally Dr. Otto Struve has found that the sequence of Cepheids fits almost perfectly on to that of newly formed binaries. Thus the prospects for the "fission theory" of Cepheid variables seem hopeful, but the theory must be very thoroughly tested before it can be accepted, and it cannot be claimed that it has been so far either tested thoroughly or accepted extensively.

An alternative view, first propounded by Plummer and Shapley, regards Cepheid variables as pulsating spheres of gas. The behaviour of such masses of gas has been investigated mathematically by Eddington and others, but it does not appear that it can be reconciled with the observed behaviour of Cepheid variables.

The Development of Binary Systems

Whatever the process of formation of binary systems may be, we experience fairly plain sailing in attempting to trace out the subsequent development of such systems. Three factors are simultaneously in operation.

Tidal Friction. The first of these three factors, which is only of brief duration, was designated "tidal friction" by Sir George Darwin, who first drew attention to it, and investigated the manner of its operation. When first a rotating mass breaks up and forms a binary system, the two components are so near that they necessarily raise tremendous tides on one another; Darwin shewed that these drive the two bodies apart, and equalise their rates of rotation in so doing. After these processes have been in operation for millions of years, the rates of rotation of the two bodies and their rate of revolution about one another must all become equal, so that each body perpetually turns the same face to its companion, and the two rotate about one another like the two masses of a dumb-bell joined by an invisible arm.

Although a sun and planet do not form a binary system in the strict technical sense, they are necessarily subject to the same forces as true binary systems. Thus we can see the operation of tidal friction in the fact that Mercury always turns the same face to the sun, and that Venus rotates so slowly on its axis that it turns the same face to the sun day after day, and probably also week after week. As we pass further out into space the effects of tidal friction rapidly diminish, but it is probably significant that the nearer planets, Earth and Mars, have days of about 24 hours each, while the remote planets Jupiter, Saturn and Uranus each have days of only about 10 hours. The period of Neptune's rotation is unknown. Apart from this we find, in a general way, that the further we recede from the sun the more rapidly the planets rotate, which is precisely the effect that ought to be produced by tidal friction.

In the same way, tidal friction has in all probability been mainly responsible for the present configuration of the earth-moon system, driving the moon away to its present distance from the earth and causing it always to turn the same face towards us. Tidal friction must of course still be in operation. The moon is responsible for the greater part of the tides raised in the oceans of the earth; these, exerting a pull on the solid earth underneath, slow down its speed of rotation, with the result that the day is continually lengthening, and will continue to do so until the earth and moon are rotating and revolving in complete unison. When, if ever, that time arrives, the earth will continually turn the same face to the moon, so that the inhabitants of one of the hemispheres of the earth will never see the moon at all, while the other side will be lighted by it every night. By this time the length of the day and the month will be identical, each being equal to about 47 of our present days. Jeffreys has calculated that this state of things is likely to be attained after about 50,000 million years.

After this, tidal friction will no longer operate in the sense of driving the moon further away from the earth. The joint effect of solar and lunar tides will be to slow down the earth's rotation still further, the moon at the same time gradually lessening its distance from the earth. When it has finally, after unthinkable ages, been dragged down to within about 12,000 miles of the earth, the tides raised by the earth in the solid body of the moon will shatter the latter into fragments (p. 234 below), which will form a system of tiny satellites revolving around the Earth in the same way as the particles of Saturn's rings revolve around Saturn, or as the asteroids revolve around the sun.

We have already noticed how the present arrangement of the earth-moon system enables us to calculate the earth's age; Jeffreys estimates that the system must have taken something of the order of 4000 million years to reach its present configuration (p. 146).

This period, which seems so long when judged by terrestrial standards, is only a moment in the life of a star. The components of the true binary star attain a configuration like that of the earth-moon system in a brief fraction of their lives, and, passing on, reach in time the configuration in which each perpetually turns the same face to the other. Up to now, tidal friction has been driving the masses ever further apart, but as soon as this stage is attained, the tides become stationary on both components, so that tidal friction goes out of operation. Thus the separation produced by tidal friction has now reached its limit, and, so far as tidal friction is concerned, the two bodies might rotate in the way just described to all eternity.

Loss of Weight. As tidal friction becomes inoperative, a new agency takes hold. We have calculated that the sun is losing weight at the rate of 250 million tons a minute, that it has been losing weight at this rate, or some comparable rate, for millions of millions of years, and will continue so to do for millions of millions of years yet to come. The earth is at its present distance from the sun because this distance is exactly suited to the present weight of the sun. If the sun's weight were suddenly reduced to half, its gravitational pull on the earth would also be reduced to half, and the earth would move to a greater distance from the sun.*

The sun's weight is not likely to be suddenly reduced to half, but it has been reduced by a thousand million tons in the last four minutes, with the result that its gravitational grip on the earth has been weakened and the earth has moved out to a wider orbit; at this moment the radius of the earth's orbit is greater than it was four minutes ago. The details can be traced out mathematically with complete precision. It appears that the earth's orbit round the sun is not a circle, or even an ellipse of small eccentricity; it is a spiral curve, like an uncoiled watch spring. Every year the earth moves a tiny step further out into the outer cold and darkness; exact calculation shews that its average distance from the sun increases at the rate of about a metre (39.37 inches) a century. The effect is of course of precisely the same kind as we have seen must be produced in the galactic system by the loss of weight of the stars. The only difference is that in the galaxy a system of thousands of millions of stars is expanding, whereas the sun-earth system consists of only two members.

* Although the details are unimportant, the actual course of events would be that the earth would begin to describe an elliptic instead of a circular orbit about the sun, the earth's *average* distance being greater than now.

Precisely similar effects must be produced by the loss of weight in the two components of a binary star. Here both components are radiating away energy, and so are simultaneously losing weight. Detailed calculation shews that they must continually recede from one another, but that the shape of their orbit will undergo no change.

Neither separately nor in combination do the two effects just described explain either the shapes or the sizes of the observed orbits of binary stars as a whole. To interpret these we must call on yet a third agency, the gravitational forces from passing stars. We have already seen how these account for the statistical distribution of orbits which is actually observed.

The combination of all three agencies, tidal friction, extending over millions of years, loss of weight extending over millions of millions of years, and disturbance from passing stars extending over a similar period, is responsible for the evolution of binary star systems. Their aggregate effect is to widen the distance between the two stars, while at the same time knocking the orbit out of shape.

Subdivision. While these changes are going on in the orbital arrangement of a binary system, the two components are themselves changing their physical condition on account of their continual loss of weight, and, as with the parent stars, this loss of weight will generally result in a shrinkage in the size of the star. The shrinkage of either component of the system causes its shape to run through the sequence of configurations we have already enumerated, and if the shrinkage continues for long enough, the component may end by further dividing into two separate masses. Either or both of the constituents of a binary system may subdivide into binary sub-systems in this way, resulting in a system of either three or four stars. H. N. Russell has shewn mathematically that when a binary system P, Q divides into a triple system, P, q, q', through Q breaking up into two constituents q, q', the distance between q and q' cannot be more than about a fifth of the original distance PQ. This theoretical law is well confirmed by observation. Fig. 13 shews a typical multiple system, and we notice that the separations in each of the various sub-systems are all quite small in comparison with those of the main systems.

FIG. 13.—A typical multiple star.

The development of the hypothetical primitive chaos has now been traced through five generations of astronomical bodies,

chaos—nebulae—stars—binary systems—sub-systems,

to which a sixth generation must be added if the stars of the sub-system happen to fission further, as, for instance, they have done in the star shewn in fig. 13. The genealogy of the stars begins with a vast tenuous nebula filling all space; the last generation consists of small, shrunken, dying stars with no capacity for further subdivision. The genealogy has been traced out primarily on theoretical grounds alone, but we need have no doubts as to its general accuracy, since observation confirms it repeatedly and at almost every step. Indeed it is hardly too much to say that the evolutionary sequence could have been discovered almost equally well from observational evidence alone, except for the hypothetical primaeval chaos, about which, from the nature of the case, observation cannot have anything to say.

The Origin of the Solar System

Almost all observed astronomical formations can be placed in the evolutionary sequence we have just discussed, either with fair certainty or with reasonable plausibility, except for one outstanding and conspicuous exception—the Solar System. Cosmogony came into being as an attempt to discover the origin of the solar system. The reasons why it limited its efforts to this particular problem are chronological; in the early days of cosmogony, astronomy was barely conscious of anything outside the solar system. The sketch just given of the findings of modern scientific cosmogony has been remarkable in that it has exhibited cosmogony taking us on a tour round the whole universe, explaining the origin and life-history of practically every object we encounter on this tour, and then becoming speechless when it is brought back home and confronted with its birthplace, the solar system.

Laplace's Nebular Hypothesis. The first serious scientific cosmogony was that embodied in the famous "Nebular Hypothesis" of Laplace. In 1755 Kant had pictured a primaeval chaos condensing into spinning nebulae, and, identifying one of these nebulae with the sun, had imagined the planets to be formed by the solidification of masses of gas shed from the nebula, much in the way in which we have supposed the stars to be born. In 1796, Laplace advanced similar ideas, which he developed in detail with a mathematical precision quite beyond the capacities of Kant. He shewed how, as its shrinkage made it spin ever faster and faster, a rotating mass of gas would flatten out, develop the lenticular form we have already discussed (fig. 3 of Plate

XVI), and then proceed to eject matter in its equatorial plane, or rather to leave it behind as the shrinkage of the main mass continued. At this stage it would look somewhat like the nebulae shewn in figs. 4 and 5 of Plate XVI, although Laplace, being unacquainted with nebulae of this type, adduced Saturn surrounded by its rings as an example of the formation to be expected at this stage (Plate XXIV, p. 234). Laplace imagined that the fringe of abandoned gas would then condense and form a single planet. As the main mass shrunk further, more gas was abandoned in the equatorial plane, which in due course condensed into another planet, and so on, until the sun left off shrinking and no more planets were born. A repetition of the same process, but on a far smaller scale, resulted in the satellites being born out of the planets.

That the hypothesis is *prima facie* plausible, is evident from its having survived, and indeed been generally accepted, for nearly a century before it encountered any serious opposition. Recently criticisms have accumulated, of so vital a nature as to make it clear that the hypothesis must be abandoned.

The sun, according to Laplace, broke up and gave birth to planets through excess of rotation. Yet both theory and observation indicate quite clearly the fate in store for a star which rotates too fast for safety; it does not found a family, but merely bursts, like an overdriven fly-wheel, into parts of nearly equal size. Spectroscopic binary and multiple systems are the relics of stars which have broken up through excess of rotation, and they do not in the least resemble the solar system.

Again, the principle of "conservation of angular momentum" requires that the rotation of the primaeval sun shall persist in the rotation of the present sun, and in the revolutions of the planets around it. On adding together the contributions from all of these, we obtain a total which ought to represent the angular momentum of the primaeval sun. In strictness a further contribution ought to be added on account of the weight of all the radiation which the sun has emitted since the planets were born. We can calculate the amount of this contribution, because we know the age of the earth with tolerable accuracy, but it proves to be entirely negligible.

The total angular momentum of the primaeval sun can be calculated with very fair accuracy, because more than 95 per cent. of the total angular momentum of the present solar system resides in the orbital motion of Jupiter. This contribution can be calculated with great exactness, so that some uncertainty in the minor contributions which make up the remaining 5 per cent. can have but little influence on the total.

When this total is calculated the startling fact emerges that the primaeval sun cannot have had enough rotation to cause break-up at all. Clearly the sun is very far from being broken up by its present rotation. Flattening of figure is the first step towards break-up, and the sun's figure is so little flattened by its present rotation that the most refined measurements have so far failed to detect any flattening at all. On adding the further angular momentum now represented in the motions of Jupiter and all the other members of the solar system, we arrive at a primaeval sun rotating about as fast as Jupiter now rotates, and shewing about the same degree of flattening of figure as Jupiter—enough to measure quite easily in a telescope, or even to detect with the eye alone but nothing like enough to cause break-up.

The sun is hardly like to have altered much since its planets were born, for the intervening 2000 million years or so represent but a minute fraction of the sun's total life. If, however, we imagine it to have shrunk appreciably in the interval, then the available amount of angular momentum would have been even more unable to break up the large primaeval sun than it is to break up the present shrunken sun. Whichever way we look at it, we reach the conclusion that the sun cannot have broken up, as Laplace imagined, through excess of rotation; indeed it can never have possessed more than a quite tiny fraction of the amount of rotation needed to break it up.

A third objection is of a somewhat different character. Laplace was a very great mathematician, and there was nothing the matter with his abstract mathematical theory, so far as it went. More refined modern analysis has confirmed it at every step, and observation does the same, as photographs of rotating nebulae (Plate XVI) bear witness. These photographs exhibit a process taking place before our eyes, which is essentially identical with that imagined by Laplace, except for a colossal difference of scale. Everything happens qualitatively as Laplace imagined, but on a scale incomparably grander than he ever dreamed of. In these photographs the primitive nebula is not a single sun in the making; it contains substance sufficient to form hundreds of millions of suns; the condensations do not form puny planets of the size of our earth, but are themselves suns; they are not eight or so in number but must be counted in millions.

We may ask why the same thing cannot happen on the smaller scale imagined by Laplace—for are not the conclusions of mathematics applicable independently of the size of the body with which we are dealing? The answer has in

CARVING OUT THE UNIVERSE

effect been given already (p. 207). Everything happens on the smaller scale according to plan until we come to the formation of the condensations; here the question of scale proves to be vital. We have seen (p. 186) how the molecules which form the sun have condensed into a star because of their great number; the molecules in a room do not condense into anything at all because they are too few. In the same way, the molecules left behind by the slow shrinkage of a sun (assuming this for the moment to rotate rapidly enough to leave molecules behind) would not condense, because at any instant there would be too few of them available for condensation. They would be shed by driblets, and a driblet of gas does not condense but scatters into space. A mathematical calculation decides the question definitely, and the decision is entirely adverse to the hypothesis of Laplace. Apart from minor details, the process imagined by Laplace explains the birth of suns out of nebulae; it cannot explain the birth of planets out of suns.

Laplace imagined his sun to be alone in space, even its nearest neighbours being too remote to influence it in any way. It was the natural supposition to make; we have already remarked how exceedingly rare an event it must be for two stars to approach near enough to influence one another. Yet no possible mode of evolution of a star which remains alone in space seems able to explain the origin of the solar system. As far back as 1750, Buffon had suggested that the solar system might have been produced through the disruption of the sun by another body. In propounding his Nebular Hypothesis, Laplace mentioned Buffon's idea, but dismissed it somewhat curtly on the grounds that it seemed unable to account for the nearly circular orbits of the planets—an ill-founded objection, as we shall soon see. Yet when we find that a single star cannot of itself give birth to a solar system, it becomes natural to investigate what happens on the rare occasions on which the evolution of a star is directed along other paths by the near approach of a second star.*

Tidal Theory. Actual collisions must be so exceedingly rare that we can leave them out of account. When two stars pass close to one another without collision, the primary effect must be that each raises tides in the other. The closer the approach, the higher the tides in general, although something must depend also on the speed with which the bodies pass one another, because this determines the length of time during which they influence one another.

It is likely that the two spiral arms which give their name and characteristic appearance to the spiral nebulae may owe their inception to a somewhat similar tidal action. Conditions here are different in that the rotation of the nebulae in any case causes them to emit matter in their equatorial planes, so that even small tidal forces should then cause this matter to concentrate in two symmetrical arms. Under stellar conditions a far closer approach is necessary to draw matter out from the star, and it is then most likely that there will be two unequal and dissimilar arms, or possibly only one arm.

If the approach is very close indeed, the tides may assume an entirely different aspect from the feeble tides which the sun and moon raise in our oceans; they may take the exaggerated forms of high mountains of matter moving over the surface of the star. An even closer approach may transform these mountains into long arms of gas drawn out from the

---

* The view of the origin of the solar system given in the present book is that generally known as the "Tidal Theory," which I propounded in 1916 as the result of a mathematical investigation as to what would happen when two stars approached close to one another in space.

body of the star. If, as will generally be the case, the two stars are of unequal weights, the lesser will in general suffer more disturbance than the weightier.

The Birth of Planets. The long arm or filament of matter drawn out of a star by tidal action is at first continuous in its structure, but analysis shews that it provides a fit subject for the operation of what we have called "Gravitational Instability." Condensations begin to form in this long arm of gas, in the way already described. As before, the smaller condensations are dissipated, while the larger increase in intensity until finally the filament breaks up into a number of detached masses—planets have been born out of the smaller star. The pairs of nebulae shewn in Plate XXII and the *lower* half of Plate XXIII are very probably under one another's tidal influence, and may serve to suggest the general nature of the process we are now considering, although it must be remembered that whatever is happening here is on an enormously greater scale than that of the solar system —if it were not, the telescope would be utterly unable to shew it to us.

When the new-born planets first begin to move as separate and independent bodies, they are acted on by the gravitational pulls of both stars, and so describe highly complicated orbits. Gradually the bigger star recedes until its gravitational effect becomes negligible, and the planets are left describing orbits around the smaller star alone. If the planets moved in a clear field of empty space, these orbits would be exact ellipses. But the great cataclysm which has just occurred must have left all sorts of débris behind. Comets, shooting-stars and other minor bodies which still survive in the solar system may represent a small part of it, but probably the main part was left in the form of dust or gas, so that the new-born planets had at first to plough their way through a medium which offered some resistance to their motions. Under these circumstances their orbits would not be strict ellipses. It can be proved that a resistance of the kind just described would change the shape of the orbits, and that with the progress of time they would become more circular, finally becoming absolutely circular if the medium should last long enough.

The débris of gas and dust would, however, continually be swept up by the planets and would disappear completely in time, probably leaving the planetary orbits something short of absolute circles. Assuming that all this has happened in the solar system, very little of the original débris can now remain, its last vestiges being probably represented by the particles of dust which are responsible for the zodiacal light. Nevertheless, the resisting medium appears to have existed for long enough to make the orbits, both of the planets and of their satellites, very nearly circular for the most part. In Jeffreys' hands, a study of the rate at which such changes would occur has yielded a valuable confirmation of other estimates of the length of time which has elapsed since the planets were born.

We may next turn our attention to the physical changes which must all this time be affecting the various planets. The long filament of matter pulled out of the sun is likely to have been richest in matter in its middle parts, these parts having been pulled out when the second star was nearest and its gravitational pull was strongest. Diagrammatically at least, we may think of this filament as shaped like a cigar—thick near the middle, thin at the ends—so that when condensations begin to form, those near the middle are likely to be richer in matter than those at the ends. This probably

PLATE XXII

Mt. Wilson Observatory.

Two Nebulae (N.G.C. 4395, 4401) suggestive of Tidal Action.

PLATE XXIII

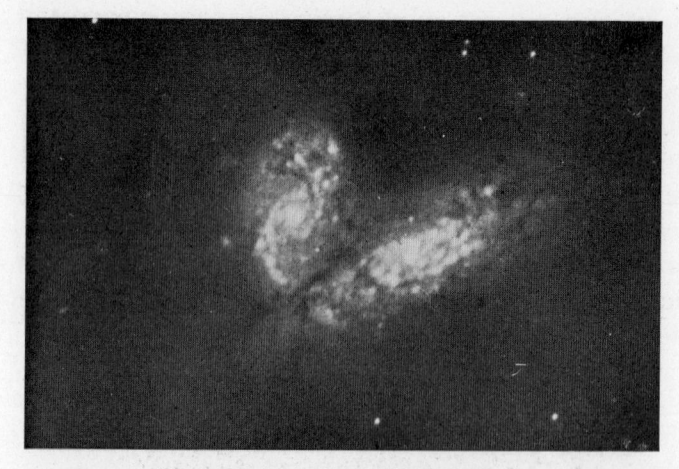

The twin Nebulae N.G.C. 4567-8.

Mt. Wilson Observatory.

The Nebula N.G.C. 7479.

explains why the two most massive planets, Jupiter and Saturn, occupy the middle positions in the sequence of planets.

Fig. 14 shews the planets arranged in the order of their distances from the sun with their sizes drawn roughly to scale. The thousands of asteroids whose orbits now fill the space between the orbits of Mars and Jupiter are represented as a single planet, it being generally supposed that these asteroids were formed by the break-up of what was originally a single planet in a way we shall shortly describe.

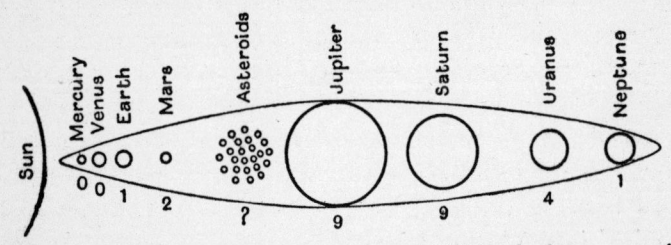

FIG. 14.—Diagrammatic scheme shewing the birth of planets out of a cigar-shaped filament of gas. The number of satellites is indicated under each planet (see p. 229).

If we surround the planets by a continuous outline, as in the diagram, we can reconstruct in imagination the cigar-shaped filament out of which they were produced, and we see at once how the biggest planets were produced where matter was most abundant.

The Birth of Satellites. We have already noticed how the great disparity of weight between the sun and planets distinguishes the sun-planet formation from that of the normal binary star, and so suggests entirely different origins for the two formations. Exactly the same disparity repeats itself in the planet-satellite systems. Just as the parent sun is enormously more massive than its children the planets, so these in turn are far more massive than their satellite children. The sun has 1047 times the weight of its most massive planet and many millions of times the weight of the smallest. In the system of Saturn the corresponding figures are 4150 and about 16,000,000. The nearest approach to equality of weights is provided by the earth-moon system, the earth having only 81 times the weight of the moon. And, like the planetary system of the sun, the satellite systems of Saturn and, to a lesser degree, of Jupiter shew a general tendency for the weights of the various satellites to increase up to a maximum as we pass outwards from the planet, and then to decrease again. This again suggests formation out of a cigar-shaped filament with matter occurring most richly near the middle. In conjunction with the repetition of the great disparity of weights between primary and secondaries, this indicates very forcibly that the satellites of the planets must have been born by the same type of process as had previously resulted in the birth of their parents.

We can imagine the process in a general way. Immediately after their birth, the planets must begin to cool down. The largest planets, Jupiter and Saturn, naturally cool most slowly and the smallest most rapidly. The latter may lose heat so speedily that they liquefy, and perhaps even solidify, almost immediately after their birth. While these events are in progress, the planets are still pursuing somewhat erratic orbits, in describing which they may pass so near to the sun that a second series of tidal disruptions occurs. In these the sun itself plays the *rôle* originally played by the passing star from space, the planets playing the part originally taken by the sun. The sun may now tear long filaments of matter out of the surfaces of the planets, and these, forming condensations, may give birth to yet another generation of astronomical bodies, the satellites of the planets. In some such way the tidal theory imagines the planetary satellites to have come into being.

Mathematical investigation shews that the more liquid a planet was at birth, the less likely it would be to be broken up by the still gaseous sun. If, however, such a break-up occurred, the weights of primary and satellites would be more nearly equal than if the planet had been more gaseous. Thus on passing from wholly gaseous planets to planets which liquefied at or immediately after their birth, we should expect at first to find planets with large numbers of relatively small satellites, and then, after passing through the borderline cases of planets with small numbers of relatively large satellites, we should expect to come to planets having no satellites at all.

We have already seen that the big central planets, Jupiter and Saturn, ought to have remained gaseous for longest and the smaller planets to have liquefied earliest; we now see that this prediction of theory exactly describes what is actually found in the solar system. Starting from Jupiter and Saturn, each with nine relatively small satellites, we pass Mars with only two satellites, and come to the earth with its one relatively large satellite, followed by Venus and Mercury which have no satellites at all. Proceeding in the other direction we leave Jupiter and Saturn each with their nine tiny satellites, to discover Uranus with four small satellites and Neptune with one comparatively big satellite. The number placed under each planet in fig. 14 gives the number of its satellites. When the numbers are exhibited in this way, the law and order in the arrangement of the satellite systems becomes very apparent, and this arrangement is seen to be exactly in accordance with the prediction of the tidal theory.

The cigar-shaped arrangement applies not only to the sizes of the planets, but also, as it ought, to the numbers of their satellites.

The earth and Neptune, with only one satellite each, and those comparatively large ones, form the obvious lines of demarcation between planets which were originally liquid and those which were originally gaseous. This leads us to conjecture that Mercury and Venus must have become liquid or solid immediately after birth, that the earth and Neptune were partly liquid and partly gaseous, and that Mars, Jupiter, Saturn and Uranus were born gaseous and remained gaseous at least until after the birth of their families of satellites.

We may perhaps find further evidence confirmatory of the tidal theory in the circumstance that the weights of Mars and Uranus are abnormally small for their positions in the sequence of planets. If, as we have supposed, the planets were all born out of a continuous filament of matter, the weight of Mars at birth would in all probability have been intermediate between those of the earth and Jupiter, and the weight of Uranus intermediate between those of Neptune and Saturn. But if, as we have already been led to suppose, the two anomalous planets Mars and Uranus were the two smallest planets to be born in the gaseous state, they would be likely to lose more of their substance than the other planets through their outermost layers of molecules dissipating away into space before they had cooled down into the liquid state. If Mars and Uranus are supposed to be mere relics of planets which were initially far more massive than they now are, the anomalies begin to disappear and the pieces of the puzzle to fit together in a very satisfactory manner.

Orbital Planes. Every rotating mass, whether gaseous,

liquid or solid, has a definite axis of rotation, and, perpendicular to this, a definite equatorial plane which divides the mass symmetrically into two exactly equal and similar halves. When a mass breaks up under its own rotation, the equatorial plane and the symmetry still persist. Illustrations of this can be found in any set of photographs of rotating nebulae, as, for instance, those shewn in Plates XV and XVI. In more humble life an illustration is provided by the splashes of mud thrown off by a spinning bicycle-wheel, which all keep in the plane in which the wheel is spinning.

If the sun's equatorial plane had proved to be a plane of symmetry for the solar system, so that the whole system was similarly arranged as regards the two sides of this plane, it might have been possible to explain the system as the result of a rotational break-up. But the sun's equatorial plane is not a plane of symmetry. The planets do not move in it, most of them moving in a plane which makes an angle of 6 or 7 degrees with it. In terms of our humble analogy, the splashes of mud are not flying about in the plane in which the bicycle-wheel is spinning.

The hypothesis that the planets came into being through a rotational break-up of the sun fails completely before this fact, but the tidal theory provides a simple explanation of it at once. The sun is still rotating much as it was before the planets were born, and so retains its original equatorial plane. The quite different plane in, or very close to, which the planets are describing orbits must clearly be the plane in which the long tidal filament was originally drawn out by the passing star. Thus the plane in which the outer planets now move must record the position of the plane in which the two stars, the sun and the wandering star, the second parent of the sun's family of children, described orbits about one another 2000 millions years ago. It is the only clue the latter has left of his identity, and is of course far too slight to make identification possible after this long lapse of time.

To sum up, we have seen that the normal mechanism by which the greater part of the universe has been carved out is the birth of successive generations of astronomical bodies through the action of "gravitational instability." The normal genealogy runs somewhat as follows:

chaos—nebulae—stars—binary systems—sub-systems.

Not all stars have passed on to the last two generations; where only a small amount of rotation was present, a star might well live its whole life without further subdivision. Our sun would have provided an instance of this had it not been for the rare accident of the close approach of a second star. From the interaction of these, two other generations came into being, still through the mechanism of gravitational instability. For our solar system, as for any other similar systems there may be in the sky, the genealogy runs as follows:

chaos—nebula—sun—planets—satellites.

Both types of genealogy shew five generations, each born from its parent through the action of gravitational instability, and between them the two genealogies include practically all the large-size astronomical objects with which we are acquainted. It is then fair to say that "gravitational instability" appears to be the agency primarily responsible for the main architecture of the universe.

Roche's Limit. The reign of gravitational instability must end with the birth of planetary satellites, since gaseous bodies of less weight than these could not hold together. Even under the most favourable circumstances their feeble gravitational pulls would be unable to restrain their outermost molecules from escaping, so that the whole mass would speedily scatter into space. Yet astronomy provides many instances of smaller bodies; we have already mentioned the asteroids,, meteors or shooting-stars, and the particles of Saturn's rings. As all these are too small to have been born in the gaseous state, we must suppose them to be the broken-up fragments of larger masses. This accords with the circumstance that these small bodies as a rule do not occur individually but in swarms.

It is common experience that shooting-stars are encountered in swarms, and, as we shall see, the motion of many of these swarms makes it possible to identify them as broken-up comets. The asteroids occur as a single swarm. If these were found scattered throughout the solar system, their origin might present a difficult problem. As things are, the whole swarm can be explained quite simply as the broken fragments of a primaeval planet. Saturn's rings again admit of a natural explanation as the fragments of a former shattered moon of Saturn. Comets, which we have hardly had occasion to mention so far, are in all probability swarms of minute bodies which are just held together sufficiently by their mutual gravitational attraction to describe a common orbit in space. At its apparition in 1909, Halley's comet was estimated to reflect as much of the sun's light as a single body 25 miles in diameter. Yet its apparent surface was 300,000 times that of such a body, and was quite transparent. It is difficult to resist the conclusion that the comet consisted of a widely-spaced swarm of small bodies, and such a swarm again admits of a simple explanation as the broken fragments of a single mass.

It is easy to imagine the mechanism by which such break-up might occur. We have supposed the sun to have been broken up, at least to the extent of ejecting a family of planets, by the tidal pull of a passing star. What would have happened if the passing star had not passed, but had come to stay? So long as it remained within a certain distance of the sun, its tidal forces were pulling the sun to pieces. We can imagine how a longer visit from it would have resulted in a greater upheaval in the sun, and the birth of a larger family of planets. Finally a visit of unlimited duration would have shattered the sun into fragments.

In 1850 Roche gave a mathematical investigation of this process of tidal break-up. His discussion dealt only with solid or liquid bodies, but the underlying mechanism is the same whether the bodies are solid, liquid or gaseous. We have seen that the smaller of the two bodies involved in a tidal encounter suffers the most. Roche dealt only with the case in which one body was very small in comparison with the other; in such a case the small body was completely broken up, while the larger one remained unscathed. Roche imagined the small body to describe an orbit of gradually decreasing size around the big body. If the two bodies were of equal density, he calculated that the small body would be broken up as soon as the radius of its orbit fell to 2.45 times the radius of the large body. If the smaller body is of lower density than the larger, the distance is correspondingly increased.

This distance is generally known as Roche's limit. A satellite can describe a circular orbit about its primary with safety so long as this orbit lies beyond Roche's limit, but it is broken into fragments as soon as it trespasses within the

PLATE XXIV

Saturn in 1916.

Saturn in 1917.

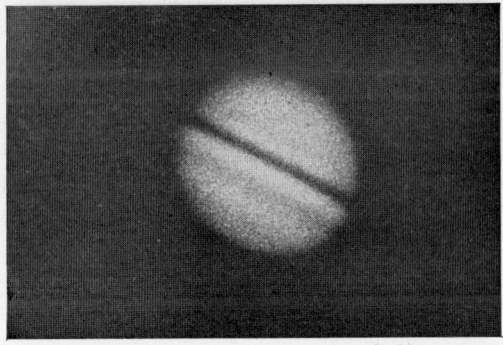

Lowell Observatory.
Saturn in 1921.
Saturn and its System of Rings.

limit. The following figures confirm Roche's mathematical analysis:

Radius of Saturn's outermost ring	2.30	radii of Saturn
Roche's limit	2.45	*radii of primary*
Radius of orbit of Saturn's innermost satellite	3.11	radii of Saturn
Radius of orbit of Jupiter's innermost satellite	2.54	radii of Jupiter
Radius of orbit of Mars' innermost satellite	2.79	radii of Mars

At the same time they suggest very forcibly that Saturn's rings are the broken-up fragments of a former satellite which ventured into the danger-zone marked out by Roche's limit. We have seen how our own moon is destined in time to contract its orbit, until it is finally drawn within the Roche's limit surrounding the earth and broken into fragments. After this the earth will have no moon, but will be surrounded by rings like Saturn.

We speak of Saturn's rings in the plural, because two distinct circular gaps cause an appearance of three detached rings. There is a tendency to jump to the hasty inference that the rings are the shattered remains of three distinct satellites, but it is not so. Goldsbrough has shewn how certain orbits around Saturn are rendered unstable by the motions of the larger satellites of Saturn, so that no particle could permanently remain in such an orbit. He has calculated positions for these unstable orbits, and these are found to agree exactly with the positions of the observed divisions between the rings. Thus Saturn's rings were in all probability produced by the breakage of a single satellite. The ring of small satellites which our moon will ultimately form round the earth will contain no divisions, because the earth has no other moons to render certain orbits unstable.

Roche's fundamental idea can be extended in many directions and admits of varied applications. There must, for instance, be a danger-zone, marked off by a Roche's limit, surrounding the sun. Comets must occasionally pass through this and become broken up in so doing. Two comets, Biela's comet (1846) Taylor's comet (1916) were observed actually to break in two while near the sun, while in 1882 a comet was seen to divide into four parts. Biela's comet returned in due course (1852) in the form of two distinct comets a million and a half miles apart, since which time neither part of the original comet has been seen again. The orbit of this comet was identical with that of the Andromedid meteors, which make a display of shooting-stars in the earth's atmosphere on favourable 27ths of November, so that it is likely that these shooting-stars are the broken remains of Biela's comet. Other conspicuous swarms of shooting-stars also move in the tracks of comets—the Leonids, which used to make a magnificent show every 33 years, move in the track of Comet 1866 I, the Perseids in the track of another Comet (1862 II), and the Aquarids in the track of Halley's famous comet. In each case, there can be little doubt that the shooting-stars are scattered fragments of the comets. Besides this there are several families of comets whose members follow one another round and round in the same orbit, as though they had originally formed a single mass.

In the same way a Roche's limit must surround the planet Jupiter, so that comets and other bodies may be broken up through getting inside the danger-zone marked off by this limit. Jupiter's innermost satellite is already perilously near it. But the greatest interest of this particular danger-zone is that it probably accounts for the existence of the asteroids. In the early days of the solar system, when the orbits of the planets were less nearly circular than they now are, a primaeval planet between Mars and Jupiter may well have described an orbit so elongated as to take it repeatedly within the danger-zone of Jupiter. If so, we need look no further for the origin of the asteroids. It is significant that the average orbit of all the asteroids agrees almost exactly with that of the planet which Bode's law (p. 19) would require to exist between Mars and Jupiter.

CHAPTER V

Stars

THE process of carving out the universe which we considered in the last chapter ends normally with a simple star, although special accidents may have other consequences. As the result of close approaches with other stars, a tiny fraction of the total number of the stars, perhaps about one star in 100,000 (p. 320 below), may be attended by a retinue of planets. Another fraction, still small, although far greater than the foregoing, appears to have broken up as the result of excessive rotation, and formed binary or perchance multiple systems. But the destiny of the majority of stars is to pursue their paths solitary through space, neither breaking up of themselves nor being broken up by other stars. The only contact such stars have with the outer universe is that they are incessantly pouring away radiation into space. This outpouring of radiation is almost entirely a one-way process, any radiation a star may receive from other stars being quite inappreciable in comparison with the amount it is itself emitting. The radiation is accompanied by a loss of weight, and this again is all give and no take, the weight of any stray matter the star may sweep up out of space, like that of any radiation it receives, being quite inappreciable in comparison with the weight it loses by radiation. Without unduly straining the facts, the normal object in the sky may be idealised as a solitary body, alone in endless space, which continually pours out radiation and receives nothing in return.

STARS

In the present chapter we shall consider the sequence of changes which such a star may be expected to experience during the course of its life. Having already discussed the mechanical accidents to which stars are liable, namely, fission through rotation and break-up through the tidal action of a passing star, we now turn to consider the life of a normal star which escapes all accidents until it finally becomes extinct through mere old age.

It will be necessary in the first place to describe the physical states of the various types of stars observed in the sky, and as a preliminary to this we must explain how the observations of the astronomer are translated into a form which gives us direct information as to the condition of the star.

Surface-Temperature. In Chapter II (p. 132) we saw how each colour of light or wave-length of radiation has a special temperature associated with it, light of this colour predominating when a body is heated up to the temperature in question. For instance, a body raised to what we call a red heat emits more red light than light of any other colour, and so looks red to the eye.

Thus if a star looks red, it is legitimate to infer that its surface is at the temperature we describe as a red-heat. If another star has the colour of the carbon of an arc-light, we may conclude that its surface is at about the same temperature as the arc. In this way we can estimate the temperatures of the surfaces of the stars.

In practice the procedure is not so crude as the foregoing description might seem to imply. The astronomer passes the light from a star through a spectroscope, thus analysing it into its different colours. By a process of exact measurement, he then determines the proportions in which the different colours of light occur. This shews at once which colour of light is most plentiful in the spectrum of the star. Either from this or from the general distribution of colours, he can deduce the temperature of the star's surface.

We have already seen (p. 116) how Planck discovered the law according to which the radiation emitted by a full radiator is distributed among the different colours or wave-

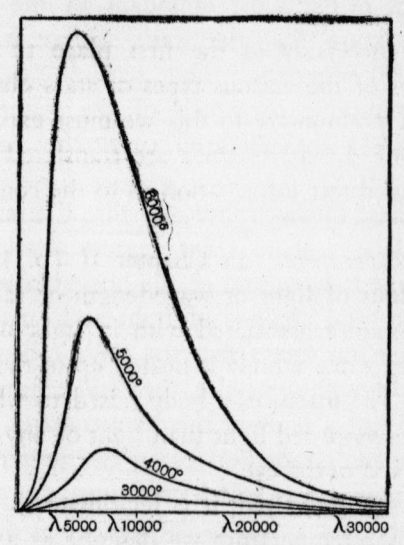

FIG. 15.—Distribution of radiation of different wave-lengths at various temperatures.

lengths of the spectrum. The four curves shewn in fig. 15 represent the theoretical distribution for the radiation emitted by surfaces at the four temperatures 3000, 4000, 5000 and 6000 degrees respectively. The different wave-lengths of light are represented by points on the horizontal axis, the marked wave-lengths being measured in the unit of a hundred-millionth part of a centimetre, which is usually called an Angstrom. The height of the curve above such a point represents the abundance of radiation of the wave-length in question.

The two methods of determining stellar temperature will be easily understood by reference to these curves. The 6000 degrees curve reaches its greatest height at a wave-length of 4800 Angstroms, so that if light of wave-length 4800 Angstroms proves to be most abundant in the spectrum of any star, we know that the star's surface has a temperature of 6000 degrees. The second method consists merely in examining to which of the theoretical curves shewn in fig. 15 the observed curve can be fitted most closely.

Either of these methods indicates that the temperature of the sun's surface is about 6000 degrees absolute, which is nearly twice the temperature of the hottest part of the electric arc. The total amount of light and heat received on earth from the sun shews that the sun's radiation must be very nearly, although not quite, the "full temperature radiation" (p. 116) of a body at this temperature. This is also shewn by the sun's radiation being distributed among the various colours in a way which conforms very closely to the theoretical curve for a full radiator at 6000 degrees shewn in fig. 15.

The surface-temperature of a star can also be estimated from its spectral type. Many of the lines in stellar spectra are emitted by atoms from which one or more electrons have been torn off by the heat of the star's atmosphere. We know the temperatures at which the electrons in question are first stripped of their atoms, and so can deduce the star's temperature.

The temperatures which correspond to the different types

of stellar spectra as shewn in Plate VIII (p. 48), are approximately as follows:

Spectral type	Temperature
B	23,000
A	11,000
F	7,400
G	6,000
K	5,100
M	3,400

The last three entries in the table refer only to normal stars having diameters comparable with that of the sun. We shall find (p. 263) that a second class of stars (giants) exist, whose diameters are enormously greater than the sun's. These have the substantially lower temperatures shewn below:

Spectral type	Temperature
G	5600
K	4200
M	3200

In studying stellar structure and mechanism, we are less concerned with the heat of the star's surface as measured by its temperature, than with the amount of radiation it pours out per square inch.

This of course depends on the temperature; the hotter a surface, the more radiation it emits. But the temperature does not measure the quantity of radiation emitted. If we double the temperature of a surface it emits 16 times, not twice, its previous amount of radiation; the radiation from each square inch of surface varies as the fourth power of the temperature. As a consequence, a star with a surface-temperature of 3000 degrees, or half that of the sun, emits only a sixteenth part as much radiation per square inch as the sun.* The radiation of each star is a compound of light, heat and ultra-violet radiation, and the proportions of these are not the same in different stars; the cooler a star's surface the greater the fraction of its radiation which is emitted as heat. Thus the star at 3000 degrees will emit nothing like as much as a sixteenth of the sun's light per square inch, but will emit more than a sixteenth of the sun's heat.

This shews that the total emission of radiation of a star cannot be estimated from its visual brightness alone; a substantial allowance must always be made for invisible radiations, both for the invisible heat at the red end of the spectrum and for the invisible ultra-violet radiation at the other end. The importance of these corrections is shewn in fig. 16. The four thick curves are identical with those already given in fig. 15, and shew how the radiation from a star of given surface-temperature is distributed over the different wave-lengths. The total radiation emitted at any temperature is of course represented by the whole area enclosed between the corresponding curve and the horizontal axis. The eye is only sensitive to radiation of wave-lengths lying between 3750 and 7500 Angstroms, so that of all this radiation only that part in the shaded strip is visible, all the rest representing invisible radiation.

We see at once that a fair proportion of the radiation emitted by a star at 6000 degrees comes within the range of visibility, but only a small fraction of that emitted by a star at 3000 degrees. Taking the stars as a whole, star-light forms only a small part of the total radiation of the stars.

* This is shown in fig. 15, the area of the 3000 degree curve being only a sixteenth of the area of the 6000 degree curve.

THE UNIVERSE AROUND US

If our eyes were suddenly to become sensitive to all kinds of radiation, and not to visual light alone, the appearance of the sky would undergo a strange metamorphosis. The red stars Betelgeux and Antares, which are at present only 12th and 16th in order of brightness, would flash out as the two brightest stars in the sky, while Sirius, at present the brightest of all, would sink to third place. A star in the

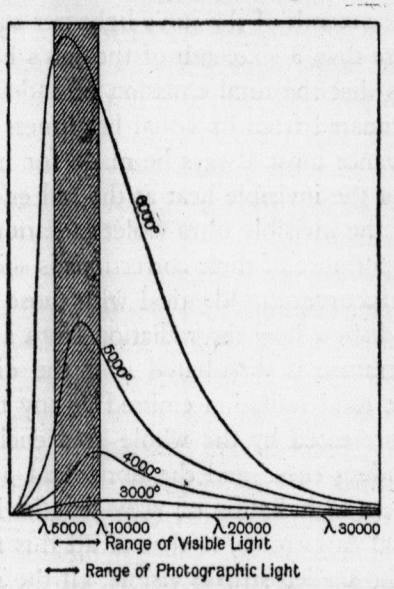

FIG. 16.—Distribution of radiation into visible and invisible.

very undistinguished constellation of Hercules would be seen as the sixth brightest star in the sky. It is the star α Herculis, at present outshone by about 250 stars. As a consequence of its extremely low temperature of 2650 degrees, this star emits its radiation almost entirely in the form of invisible heat. For instance it emits 60 times as much heat as the blue star η Aurigae, whose temperature is about 20,000 degrees, but only four-fifths as much light.

Allowances for invisible radiation have been made in all the calculations referred to in the present book, although it has not been thought necessary continually to restate this.

Stellar Diameters. It is easy to measure the diameter of a planet, because this appears in the telescope as a disc of finite size. But the stars are too remote for their diameters to be measured in the same way. No star appears larger in the sky than a pin-head held at a distance of six miles, and no telescope yet built can shew an object of this size as a disc. All stars, even the nearest and largest, appear as mere points of light, so that their diameters can only be measured by roundabout methods.

When a star's distance is known, we can tell its luminosity from its apparent brightness. From this, after allowing for invisible radiation, we can deduce the star's total outpouring of energy—so many million million million million horsepower. We also know its outpouring of energy per square inch of surface, because this depends only on its surface-temperature which we deduce directly from spectroscopic observation. Knowing these two data, it is a mere matter of simple division to calculate the number of square inches which make up the star's surface, and this immediately tells us the diameter of the star.

The diameters of exceptionally large stars may be measured more directly by an instrument known as the Interferometer. When we focus a telescope on a star we do not, strictly speaking, see only a point of light, but a point of light surrounded by a rather elaborate system of rings of alternating light and darkness, called a diffraction pattern. It might be thought that the size of these rings would tell us the size of the star, but the two have nothing to do with one another. The rings represent a mere instrumental defect, their size depending solely on the size and optical arrangement of the telescope. Following a method suggested by Fizeau in 1868, Professor Michelson has shewn how even this defect can be turned to useful ends, and by its aid has produced what is perhaps the most ingenious and sensational instrument in the service of modern astronomy—the interferometer. In effect, this instrument superposes two separate diffraction patterns of the same star, and sets one off against the other in such a way as to disclose the size of star producing them. The diameter of a few of the largest stars have been measured in this way, so that we may say that we know their sizes from direct observation. In every case the directly measured diameter agrees fairly well, although not perfectly, with that calculated indirectly in the way already explained. The discrepancies, which are not of serious amount, appear to result from red stars not being accurate "full radiators" in the sense explained on p. 116.

The interferometer method is only available for the largest stars, but at the extreme other end of the scale the theory of relativity has come to the rescue. Einstein shewed it to be a necessary consequence of his theory of relativity that the spectrum of a star should be shifted towards the red end by an amount depending on both the weight and the diameter of the star. If, then, a star's weight is known, the observed spectral shift ought immediately to tell us its diameter. This spectral shift has recently been observed in the light received from the companion of Sirius, and measurements of its amount lead to a value for the star's diameter which agrees exactly with that calculated from its luminosity. Thus at both ends of the scale, for the very largest as well as for the very smallest of stars, direct observation confirms the values calculated for the diameters of the stars.

We may accordingly feel every confidence in the calculated diameters of all stars, even when these cannot be checked by direct measurement. Indeed a discrepancy between the true and calculated diameters could only arise in one way. The diameters are calculated on the assumption that the stars emit their full temperature-radiation. If the stars had been partially transparent like the nebulae, or solid bodies like the moon, this assumption would have been false, and its falsity would at once have been shewn by discordances between the calculated and measured diameters of the stars. The fact that no large discordances appear suggests that the stars emit nearly full temperature-radiation throughout the whole range of size from the largest to the smallest.

The Variety of Stars

Observation shews that the physical characteristics of the stars vary enormously, so that it is easy, as we shall soon see, to tell a sensational story by contrasting extremes, setting the brightest against the dimmest, the biggest against the smallest, and so on. This would, however, give a very unfair impression of the inhabitants of the sky; it would be like judging a nation from the giants and dwarfs, the strong men and the fasting men, seen inside the showman's tent.

We shall obtain a more balanced impression of the actual degree of diversity shewn by the stars as a whole if we consider the physical states of those stars which are nearest the sun. By taking these precisely in the order in which they come, we avoid any suspicion of going out of our way to introduce stars merely because they are bizarre or exceptional. The small group of stars obtained in this way may be expected to form a fair sample of the stars in the sky, although of course it will not be a large enough sample to include extremes. We need not discuss the sun itself in detail because it will figure as our standard star, with reference to which all comparisons are made.

The System of α Centauri. This system consists of three constituent stars, which are believed to be our three nearest neighbours in space.

The brightest, α Centauri *A,* is very similar to the sun. It is of the same colour and spectral type, but weighs 14 per cent. more and is about 12 per cent. more luminous. Being of the same colour as the sun, it emits the same amount of radiation per square inch. Thus its 12 per cent. greater luminosity shews that it must have a surface 12 per cent. greater, and therefore a diameter 6 per cent. greater, than the sun.

The second constituent, α Centauri *B,* is considerably redder than the sun, its surface-temperature being only about 4400 degrees against the sun's 6000 degrees or so. It has 97 per cent. of the sun's weight, but only about a third of its luminosity. Yet, as a consequence of its low temperature, it needs 50 per cent. more area than the sun to discharge a third of the sun's radiation; this makes its diameter 22 per cent. greater than that of the sun. α Centauri *A* and α Centauri *B* together form a visual binary, the two components revolving about one another in a period of 79 years.

Neither of these two constituents is very dissimilar from the sun, but the third star of the system, Proxima Centauri, is of an altogether different type. It is red in colour, with a surface-temperature of only about 3000 degrees. It is exceedingly dim, emitting only a ten-thousandth part as much light as the sun, and has only a fourteenth part of the sun's diameter. Its weight is unknown.

The sizes of the three stars of this system, with that of the sun for comparison, are shewn in fig. 17.

Munich 15040. This is a single faint star about which little is known. Its surface is red, with a temperature probably little above 2500 degrees,, and it emits only 1/2500th of the light of the sun.

Wolf 359. This is the faintest star yet discovered, but beyond this very little is known about it. It is red in color and emits only about 1/50,000th of the light of the sun.

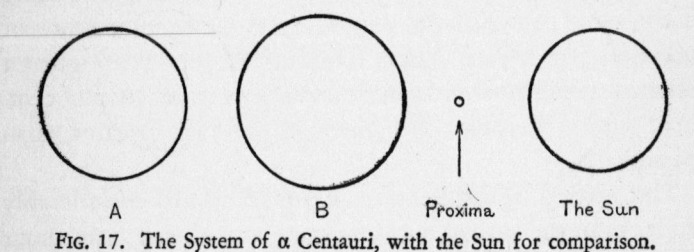

FIG. 17. The System of α Centauri, with the Sun for comparison.

Lalande 21185. Another faint red star, emitting 1/200th of the light of the sun.

The System of Sirius. This consists of two very dissimilar stars, there being some suspicion that a third also may exist.

The principal star, Sirius *A*, which appears as the brightest star in the sky (the Dog-star), is white in colour and has a surface-temperature of about 11,000 degrees. As this is nearly twice the sun's temperature, Sirius *A* emits nearly 16 times as much radiation per square inch as the sun. Its luminosity is about 26 times that of the sun, and this requires the star's diameter to be 58 per cent. greater than that of the sun. It has nearly four times the sun's volume, but only 2.45 times its weight, so that matter is not as closely packed in Sirius *A* as in the sun. An average cubic metre contains 1.42 tons in the sun, but only 0.93 ton in Sirius *A*.

The faint companion Sirius *B* is one of the most interesting stars in the sky. It is of nearly the same colour and spectral type as Sirius *A*, but emits only a ten-thousandth part as much light. After allowing for the slight difference in surface-temperature, we find that its surface is only 1/2500th, and its diameter 1/50th of that of Sirius *A*. Yet

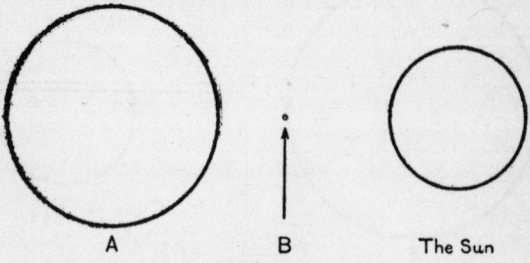

Fig. 18.—The System of Sirius, with the Sun for comparison.

Sirius *A* weighs only three times as much as Sirius *B*, although having 125,000 times its volume. It is not Sirius *A* but Sirius *B* that is remarkable; the average density of matter in the latter is about 60,000 times that of water, the average cubic inch containing nearly a ton of matter. Fig. 18 shews the sizes of the two components of Sirius drawn to the same scale as fig. 17.

B.D. 12° 4523 *and Innes* 11 h. 12 m. 57.2°. Two stars, as to the physical state of which nothing is known except that they are very faint, emitting 1/1400th and 1/10,000th of the sun's light respectively.

Cordoba, 5 h. 343 and τ *Ceti.* Two faint stars, both of

reddish colour, emitting 1/600th and a third of the sun's light respectively.

The System of Procyon. This is a binary system, similar in many respects to Sirius. The main star, Procyon *A*, is of the same general type as the sun, but weighs 24 per cent. more, and emits 5½ times as much light. Its surface-temperature is about 7000 degrees, and its diameter 1.80 times that of the sun.

The faint companion, Procyon *B*, is so faint that nothing

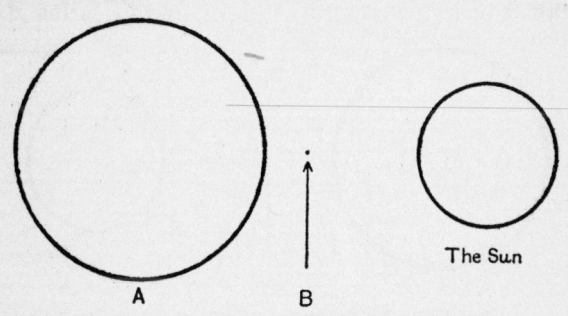

FIG. 19.—The System of Procyon, with the Sun for comparison.

is known as to its physical condition except that it emits only 1/30,000th of the light of the sun. Its weight is 39 per cent. of the sun's weight.

Fig. 19 shews the sizes of the two components of Procyon on the same scale as before.

Next in order, as we recede from the sun, come eight very undistinguished stars, every one of which is redder and fainter than the sun, none of them having a surface-temperature higher than 5000 degrees, and none of them emitting more than a quarter of the sun's light. After these we come to

THE UNIVERSE AROUND US

The System of Kruger 60. This is a binary system in which both components are small, red and dim.

The brighter component Kruger 60 *A*, has a surface-temperature of 3200 degrees, and emits 1/400th of the light of the sun. Its diameter is a third, and its weight a quarter of the sun's, so that its substance must be packed about 7 times as closely as that of the sun.

The fainter component Kruger 60 *B*, has a similar surface-temperature but emits only 1/14,000th of the sun's light. Its diameter is a sixth and its weight a fifth of the sun's;

FIG. 20.—The System of Kruger 60, with the Sun for comparison.

so that its substance must be packed about 40 times as closely as the sun's. The system is illustrated in fig. 20.

van Maanen's star. Another very faint star, which has the high surface-temperature of 7000 degrees. Notwithstanding this, it only emits 1/6000th of the sun's light. Consequently its diameter is only about 1/110th of the sun's, the star being if anything smaller than the earth. Its weight is unknown, but its substance is in all probability packed even more closely than in Sirius *B*.

The discussion of this sample of stars suggests that the majority of stars in space are smaller, cooler and fainter than the sun. Stars exist which are far brighter than the sun, but they are exceptional, the average star in the sky being a small dull dim affair in comparison with our sun.

STARS

With this sample of the average population of the sky before us, we may proceed to discuss the various characteristics of stars in a systematic way, without fearing to mention extremes. Let us begin with their weights.

Stellar Weights. The two stars of smallest known weight in the whole sky are the faint constituent of Kruger 60, just discussed, and the faintest constituent of the triple system o_2 Eridani, each of which has a fifth of the sun's weight. But the stars whose weights are known are so few that there can be no justification for supposing these to be the smallest weights which occur in the whole universe of stars. A general survey of the situation, on lines to be indicated later (p. 265), suggests that there may be many stars of still smaller weight, but that very few are likely to have weights which are enormously smaller. Probably very few stars weigh as little as a tenth of the sun's weight.

The vast majority of stars have weights intermediate between this and ten times the sun's weight. Stars which weigh even three times as much as the sun are rare, those which weigh ten times as much are very rare, probably only about one star in 100,000 having ten times the weight of the sun. Even higher weights undoubtedly occur—we have already mentioned Plaskett's star, whose two constituents have more than 75 and 63 times the sun's weight respectively, and the quadruple system 27 Canis Majoris which to all appearances weighs 940 times as much as the sun—but such instances are very, very unusual. We may say that as a general rule the weights of the stars lie within the range of from a tenth to ten times the sun's weight, and we shall find that stars differ less in their weights than in most of their other physical characteristics.

Luminosity. A far greater range is shewn, for instance,

in the luminosities of the stars—in their candle-powers measured in terms of the sun's candle-power as unity. The most luminous star known is *S* Doradus, already mentioned, with 300,000 times the luminosity of the sun, while the least luminous is Wolf 359 with only a fifty-thousandth part of the luminosity of the sun. The range of stellar luminosities, as of stellar weights, extends about equally on the two sides of the sun, so that the sun is rather a medium star in respect both of weight and luminosity. It is medium in the sense of being about half-way between extremes, but we have seen that there are many more stars below than above it.

In comparison with the very moderate range of stellar weights, the range of luminosity is enormous; *S* Doradus is 15,000,000,000 times as luminous as Wolf 359. If *S* Doradus is a lighthouse, Wolf 359 is something less than a firefly, the sun being an ordinary candle. If the sun suddenly started to emit as much light and heat as *S* Doradus, the temperature of the earth and everything on it would run up to about 7000 degrees, so that both we and the solid earth would disappear into a cloud of vapour. On the other hand, if the sun's emission of light and heat were suddenly to sink to that of Wolf 359, people at the earth's equator would find that their new sun only gave as much light and heat at mid-day as a coal fire a mile away; we should all be frozen solid, while the earth's atmosphere would surround us in the form of an ocean of liquid air. So far as we know, there is no possibility of the sun suddenly beginning to behave like *S* Doradus, but we shall see later that the possibility of its behaving like Wolf 359 is not altogether a visionary dream.

Surface-temperature and Radiation. Sirius has been found to have the highest surface-temperature of all the stars near the sun; it is about 11,000 degrees, or nearly double that of the sun. Going further afield, we find many stars with far higher surface-temperatures. For instance, Plaskett's star is credited with a temperature of 28,000 degrees, although it must be admitted that a substantial element of uncertainty enters into all estimates of very high stellar temperatures.

At the other extreme, stellar temperatures ranging down to about 2500 degrees are comparatively common. The lowest temperatures of all are confined to variable stars of a very special type (long period variables) in which the light-variation is accompanied by, and indeed mainly arises from, a variation in the temperature of the star's surface. The temperature of these stars when at the lowest, ranges down to 1650 degrees, which is but little above the temperature of an ordinary coal fire. In many of them, the temperature varies through a large range, but it never sinks so low that the star becomes completely invisible. Thus there is a range of temperature below about 2500 degrees which no star is known to occupy, except for the long-period variables which only enter it at intervals. This would seem to suggest that the number of absolutely dark stars in the sky is quite small. Other lines of evidence lead to the same conclusion. If a star ceased to shine, its gravitational pull would still betray its existence. Although we could not detect a single dark star in this way, we could detect a multitude. If two stars out of three were dark, we should probably suspect the existence of the dark stars from their effects on the motions of the remainder so that general gravitational considerations preclude the possibility of there being a great number of dark stars.

So far as our present knowledge goes, the temperature

of stellar surfaces ranges, in the main, from about 30,000 degrees down to about 2500, the lower limit being extended to about 1650 for long-period variables at their lowest temperatures.

Apart from the long-period variables, this is only a 12 to 1 range, so that the temperatures of the stars are more uniform than either their luminosities or their weights. We must, however, remember that a star's radiation per square inch is far more fundamental than its surface-temperature, and that a 12 to 1 range in the latter involves a range of over 20,000 to 1 in the former. If we include the long-period variables, there is a range of about 110,000 to 1 in the emission of radiation per square inch.

In terms of horse-power, the sun emits energy at the rate of 50 horse-power per square inch, a star with a surface-temperature of 1650 degrees emits only 0.35 horse-power per square inch, while Plaskett's star, with a surface-temperature of 28,000 degrees, emits about 28,000 horse-power per square inch. In plain English, each square inch of this last star pours out enough energy to keep an Atlantic liner going at full speed, hour after hour, and century after century. And the energy emitted per square inch by the surfaces of various stars covers the whole range from the power of a liner to that of a man in a row-boat.

Size. The four stars of largest known diameter are the following:

Star	Diameter in terms of sun	Diameter in miles
Antares	450	390,000,000
a Herculis	about 400	346,000,000
o Ceti (at max.)	300	260,000,000
Betelgeux (average)	250	216,000,000

STARS

All these diameters have been measured directly by the interferometer. On the scale used in figs. 17 to 20, in which the sun is about the size of a sixpence, the circle necessary to represent o Ceti would be as large as the floor of a good-sized room, while the second star of the system (for o Ceti is binary) would be the size of a grain of sand. We may obtain some idea of the immense size of these stars by noticing that every one of their diameters is larger than the diameter of the earth's orbit, so that if the sun were to expand to the size of any one of them we should find ourselves inside it.

These stars must be exceedingly tenuous. Antares, for instance, occupies 90,000,000 times as much space as the sun, so that if its substance were as closely packed, it would weigh 90,000,000 times as much as the sun. Yet, in actual fact, it probably has only about 40 or 50 times the sun's weight, the difference between this number and 90,000,000 arising from the difference between the densities of Antares and the sun. On the average a ton of matter in the sun occupies considerably less than a cubic yard; in Antares it occupies rather more space than the interior of Saint Paul's Cathedral. Yet a detailed study of stellar interiors shews that we can attach but little meaning to an average of this sort. It is quite likely that matter at the centre of Antares is packed nearly (but not quite) as closely as matter at the centre of the sun (p. 275 below). The huge size of Antares is probably due mainly to an enormously extended atmosphere of very tenuous gas, and there is not much point in striking an average between this and the compact matter at the centre of the star.

The mysterious objects known as planetary nebulae, of which examples are shewn in Plate II, p. 28, ought perhaps to be regarded as stars of still larger diameter. At the centre of each of these the telescope discloses a comparatively faint star with an extremely high surface-temperature. Surrounding this is the nebulosity from which these objects derive their somewhat unfortunate name. This is in all probability merely an atmosphere of even greater extent than that surrounding the four stars of our table. Van Maanen estimates the diameter of the nebulosity of the Ring Nebula in Lyra (fig. 2 of Plate II) to be 570 times that of the earth's orbit, or about 106,000,000,000 miles. This nebulosity, however, differs from the atmosphere of an ordinary star in being very nearly transparent; we can see through 106,000,000,000 miles of the Ring Nebula but can only see a few tens or hundreds of miles into an ordinary star.

At the other extreme of size, the smallest known star, van Maanen's star (p. 252) is just about as large as the earth; over a million such stars could be packed inside the sun and still leave room to spare. And yet its weight is in all probability comparable, not with that of the earth, but with that of the sun; at a guess it may have about a fifth of the sun's weight. To pack a fifth of the sun's substance inside a globe of the size of the earth, the average ton of matter must be packed into a space of about the size of a small cherry—ten tons or so to the cubic inch. The solidity of the earth suggests that its atoms must be packed pretty closely together, but the atoms in van Maanen's star must be packed 66,000 times more closely.

How is it done? As we shall shortly see, there is only one possible answer. The atom consists mostly of emptiness—we compared the carbon atom to six wasps buzzing about in Waterloo Station. Let us break the atom up into its constituent parts, pack these together as closely as they will go, and we see the way in which matter is packed in van Maanen's star. Six wasps which can roam throughout the whole of Waterloo Station can nevertheless be packed inside a very small box.

Giants and Dwarfs. There is a continuous series of stars between the limits of weight we have mentioned, and the same is true of the limits of temperature (and so also of colour), and of size.

Within these specified limits, I can find you a star of any weight or of any colour or of any size you like. But this does not mean that you may specify the weight *and* colour *and* size of the star you want, and that I will undertake to find it for you; if the weight is right the colour may be wrong, and so on. For instance, if you ask for a red star I can find you a very heavy one or a very light one, but it is no good your asking for one of intermediate weight. So far as we know, red stars of intermediate weight simply do not exist. The same is true as regards size—there are no red stars of intermediate size. Hertzsprung noticed in 1905 that the red stars could be divided sharply into two distinct classes characterised by large and small size—he called them Giants and Dwarfs. Russell, studying the question further in 1913, confirmed Hertzsprung's earlier conclusions, and shewed that the giant-dwarf division extended to stars of other colours than red.

Imagine that we have a series of coloured ladders, one for each colour of star—red, orange, etc. Take all the red stars and stand them (in imagination) on the different rungs of the red ladder. Do not merely place them on at random; arrange them in order of their luminosities, placing those of highest luminosity uppermost. Further let several stars stand on the same rung if their luminosities are about equal.

THE UNIVERSE AROUND US

To make the arrangement definite, let each rung of the ladder represent 5 times higher luminosity than the rung immediately below it, so that each rung has a definite luminosity associated with it.*

With this agreement we are now ready to proceed. We take our red stars and place each on the appropriate rung of the red ladder, and so on for each other colour. The result is shewn diagrammatically in fig. 21, the different stars being represented by crosses.

The red stars will be found to lie as on the right of the diagram, Hertzsprung's division into giants and dwarfs being very clearly marked. The orange stars lie as on the next ladder to the left; as Russell found, the division again appears, but is less marked.

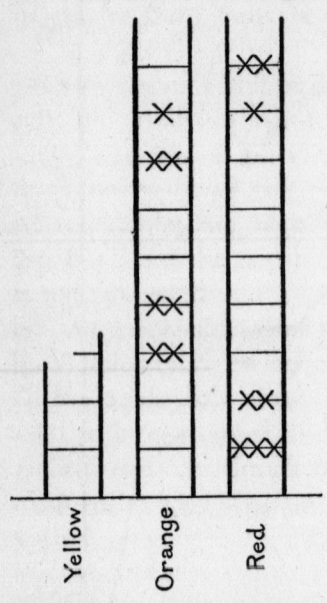

FIG. 21.—Stars of different colours arranged in order of luminosity.

The Russell Diagram. Let us make ladder diagrams of this kind for each colour of star, and put them side by side in their proper order, so as to represent stars of all possible colours. We obtain a diagram of the kind shewn in fig. 22. This type of diagram was

---

* For purely practical reasons the height is not taken proportional to the luminosity but to its logarithm; without some such device as this it would be impossible to represent the range of more than 1,000,000 to 1 in the observed luminosities of red stars.

[260]

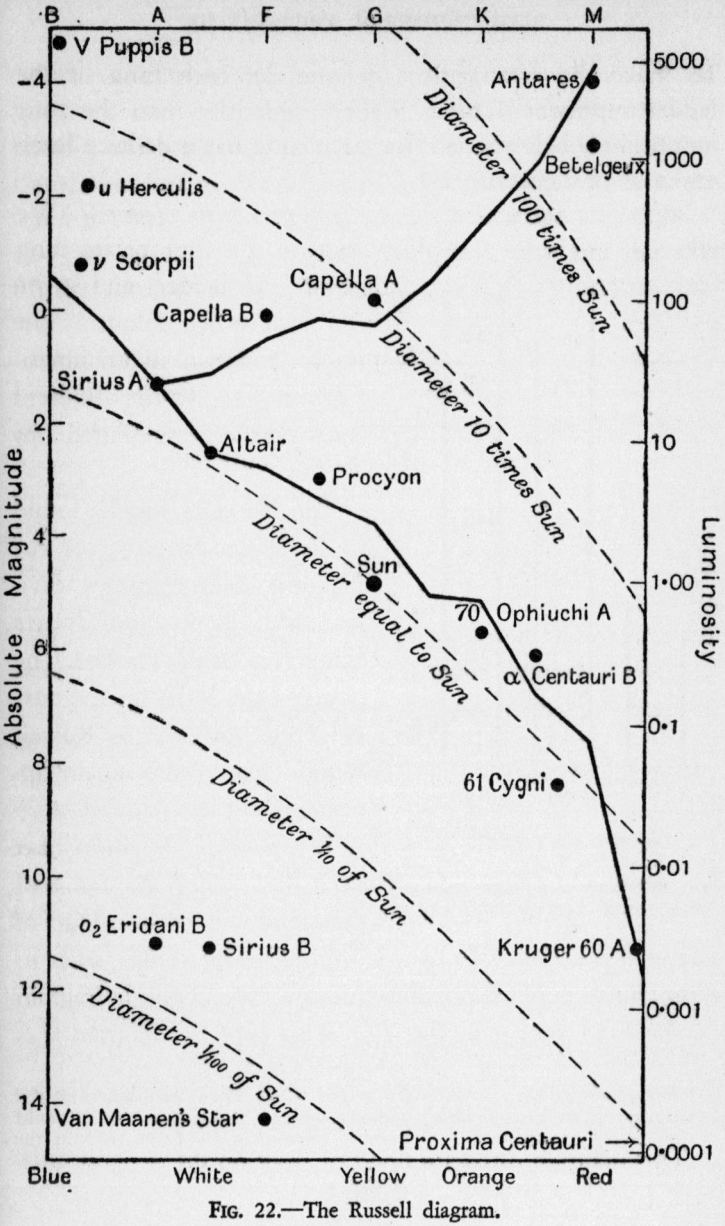

FIG. 22.—The Russell diagram.

introduced by Russell in 1913, and is now generally known as a Russell diagram.

The letters at the top of the diagram represent spectral types of stars, because these provide a better and more exact working classification than the names of colours. The colours which approximately correspond to the various spectral types are indicated at the bottom of the diagram.

Only a very few sample stars are shewn, but all known stars are found to be concentrated around the positions of these few typical stars. Broadly speaking, there are two distinct and disconnected regions which are occupied by stars. First, and most important, is a region shaped rather like a reversed γ: the central line of this region is marked in by a continuous thick line, following a determination of its position by Redman. Second, there is a smaller region near the left-hand bottom corner of the diagram. The stars which occupy this region are very faint, and have far higher surface-temperatures than other stars of similar luminosity.

We have already seen how a star's diameter can be calculated from its surface-temperature and luminosity. This amounts to the same thing as saying that two stars which occupy the same position in the Russell diagram, must have the same diameter. Thus there is a definite diameter associated with each point in the diagram, and we can map out stellar diameters in the diagram, just as we can map out heights above sea-level on a geographical map, by a system of "contour lines." In the present case the "contour lines" prove to be a system of almost parallel curves. These lie roughly as shewn by the broken lines in fig. 22, all stars lying on any one line having the same diameter.

This diagram throws a flood of light on the general question of stellar diameters. We see at once that stars of the biggest diameters—100 times the sun's diameter or more—must necessarily be red stars of high luminosity. And in actual fact the stars of large diameter shewn in the table on p. 256 are all red and have very high luminosities; they are red giants.

The majority of the stars in the sky lie in the belt which runs across the diagram of fig. 22 from top left-hand to bottom right-hand. This is known as the "main-sequence." The position of this band with reference to the "contour lines" of diameters shews that main-sequence stars are of moderate diameters. The brightest of all may have twenty times the diameter of the sun, while the faintest may have only about a twentieth of the sun's diameter, but they all have diameters which are at least comparable with that of the sun. The sample of stars from near the sun, which we have already discussed, provides many instances of main-sequence stars; we have, in order of decreasing luminosity:

Star	Luminosity	Diameter (in terms of sun)
Sirius A	26.3	1.58
Procyon A	5.5	1.80
a Centauri A	1.12	1.07
Sun	1.00	1.00
a Centauri B	0.32	1.22
τ Ceti	0.32	0.95
ϵ Indi	0.15	0.82
Kruger 60 A	0.0026	0.33
" B	0.0007	0.17
Wolf 359	0.00002	0.03

This table shews clearly how stellar luminosity and diameter decrease together as we pass down the main-sequence.

The remaining group of stars in fig. 22, those in the

bottom left-hand corner, are generally known as "white dwarfs." Their position in the diagram shews that their diameters must be excessively small. The vicinity of the sun provides three examples of this class of star, as shewn in the following table:

Star	Luminosity	Diameter (in terms of sun)
Sirius B	0.0026	0.03
o_1 Eridani B	0.0031	0.018
van Maanen's star	0.00016	0.009

In addition to these the faint companion of o Ceti is certainly a white dwarf, while Procyon B may be. These are the only known examples of white dwarfs, but the extreme faintness of these stars makes them very difficult of detection, so that it is quite likely that they are frequent objects in space.

In the table opposite, the main-sequence stars were intended to be arranged in the order of luminosity, but this happens also to be the order of weights. The weights of three of the stars are unknown; those of the remainder are as follows:

Star	Luminosity	Weight (in terms of sun)
Sirius A	26.3	2.45
Procyon A	5.5	1.24
a Centauri A	1.12	1.14
Sun	1.00	1.00
a Centauri B	0.32	0.97
Kruger 60 A	0.0026	0.25
Kruger 60 B	0.0007	0.20

STARS

Like the luminosities, the weights fall off steadily as we pass down the main-sequence, although, as already remarked, weight falls far less rapidly than luminosity.

The only stars whose weights can be measured directly are the components of binary systems, and these are relatively few in number. Seares found, however, that the weights of binary systems conformed to the law of equipartition of energy already explained in Chapter III, so that it is highly probable that other stars which are not binary also conform, for it is difficult to imagine any reason why binary systems should attain to a state of equipartition sooner than other stars. It will be remembered that this state is defined by a purely statistical law connecting the weights and speeds of motion of stars, so that the fact that a system of stars has attained this state can give no information as to the weight of an individual star whose speed is known, but it makes it possible to determine the average weight of any group of stars in terms of their average speeds of motion. In this way Seares has determined the average weights of stars of different assigned luminosities and spectral types—in other words, the average weights of the stars represented at the various points in the diagram of fig. 22. The results he obtained are shewn by the thick curved lines in fig. 23. The arrangement of these curves confirms the inference we have drawn from a few selected stars; the weight of main-sequence stars falls off steadily as we pass down the sequence from high luminosity to low.

These curved lines specify the average weight of the stars represented at each point in the Russell diagram, and the diameters are already known from fig. 22. From these two data the mean density of the star can of course be calculated. The mean densities as calculated by Seares are shewn by the broken lines in fig. 23.

This completes our collection of observational material. We now turn to the far more difficult problem of discussing what it all means. Here we leave the firm ground of ascer-

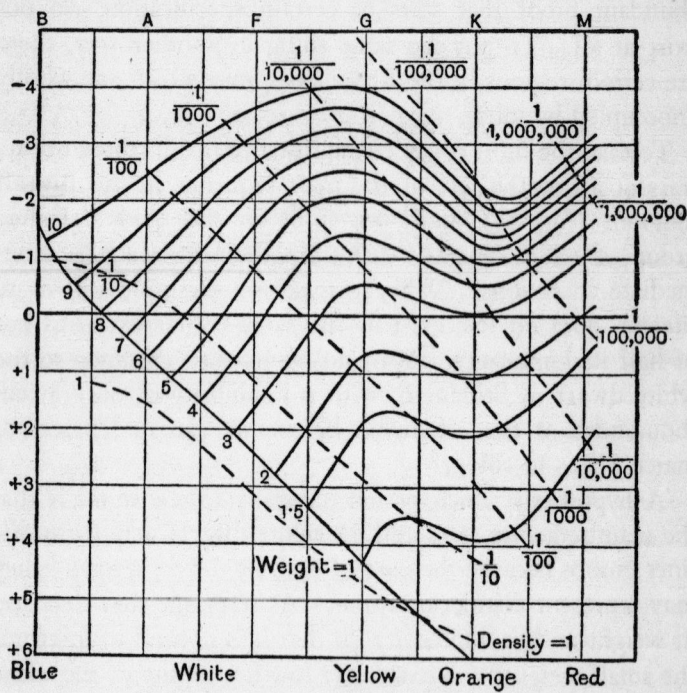

FIG. 23.—Stellar weights and densities in the Russell diagram, according to Seares.

tained fact, to enter the shadowy morasses of conjecture, hypothesis and speculation. The questions we shall discuss are some of the most interesting in the whole of astronomy, to which it must be admitted that science has so far obtained only lamentably dusty answers. The reader who is hot for certainties may prefer to read something other than the remainder of the present chapter.

The Physical Condition of the Stars

The foregoing collection of observational data has provided abundant proof that stars to certain specifications do not exist at all. To put the same thing in another way, there are certain regions in the Russell diagram which are wholly unoccupied by stars.

To take the most conspicuous instance of all, there are no stars at all to the left of the main-sequence in the Russell diagram (fig. 22), until we come to the quite detached group of white dwarfs. Why are there no stars in intermediate conditions? Why, to make the example still more precise, does no star exist of the same colour as Sirius but of half its luminosity? Why do we have to go down to the white dwarf o_2 Eridani *B,* with a luminosity of only a ten-thousandth of that of Sirius, before we can find a star to match Sirius in colour?

A hypothesis which occurs naturally to the mind is that the main-sequence stars and the white dwarfs may form distinct groups because they are of entirely different ages—they may represent distinct creations. As stars age they decrease in weight and in luminosity, so that it is natural to interpret the small weights and extremely low luminosity of the white dwarfs as evidence of an age far greater than the age of the normal main-sequence stars. Yet this hypothesis does not appear to be tenable.

With the single exception of van Maanen's star, every star which is either known or suspected to be a white dwarf forms one component of a binary system, and in every case its companion is a main-sequence star or (in the case of o Ceti) a red giant. We have already seen how rare it is for two stars to approach near to one another in space. It must be an almost inconceivably rare event for two stars, originally moving as independent bodies, so to meet in their random wanderings, that the big one "captures" the little one, and they henceforth journey together through space. For it can be shewn that, for such an event to occur, something more than a close approach is needed; a close approach must take place in the presence of yet a third star, so that no fewer than three stars must chance to come near one another simultaneously in their wanderings through the vast emptinesses of space. It is almost inconceivable that this should happen in a single instance, but it is straining the probabilities too far to suppose that it has happened in the case of every single known white dwarf but one. Thus we have to suppose that the white dwarfs and their more normal companions have been together since birth, and so were born at the same time out of the same nebula.

The difference between white dwarfs and main-sequence stars cannot, then, be a mere difference of age, and it would seem as though there must be some physical reason militating against the existence of stars in intermediate conditions. Taking a more general view of the question, we are led to investigate whether the absence of stars built to certain specifications can be attributed to such stars needing physical properties which nature cannot provide. This leads directly into the general question of the structure and mechanism of the stars.

The Internal Constitution of the Stars

Most investigations on the structure of the stars have proceeded on the supposition that their interiors are gaseous throughout. Without accepting this supposition as final truth, we may adopt it for the moment, for the purely opportunist reason that it provides the most convenient line of approach to an excessively difficult problem.

A mathematical theorem, generally known as Poincaré's theorem, proves to be of the utmost service in discussing the internal state of a gaseous star. We have seen how Helmholtz thought that the energy of the sun's radiation might come from the sun's contraction, each layer falling in upon the next inner layer as the latter shrunk, and transforming the energy set free by its fall into heat and light. It is easy to estimate how much energy would be set free by a contraction of this kind. For instance, Lord Kelvin calculated that the contraction of the sun, as it shrunk from infinite size to its present diameter of 865,000 miles, would liberate about as much energy as the sun now radiates in 50 million years. In terms of ergs, the sun's shrinkage would liberate 6×10^{48} ergs of energy.

Poincaré's theorem states that the total energy of motion of all the molecules in any gaseous star whatever is equal to precisely half the total energy which the star would have liberated in shrinking down to its present size. The theorem is true quite independently of whether the star ever has so shrunk or not: nothing is involved but the present state of the star.

One interesting consequence is that the further a gaseous star shrinks, the hotter it becomes; if a star shrinks to half its present size, the total energy set free by its shrinkage from infinite size is doubled, so that the total energy of motion of its molecules is doubled, and therefore its average temperature is doubled. This is a special case of what is generally known as Lane's Law.

Let us go on with our calculation for the special case of the sun. Poincaré's theorem tells us that, if the sun is gaseous, the total energy of motion of all its molecules is 3×10^{48} ergs. The next thing we want to know is how many molecules there are in the sun. The sun's weight is 2×10^{33} grammes, but how many molecules are there to a gramme? The answer of course depends on the type of molecule concerned; there are 3×10^{23} molecules in a gramme of hydrogen, 2×10^{22} in a gramme of air and only 2.5×10^{21} in a gramme of uranium.

If we suppose the sun to be made of air, it must consist of 4×10^{55} molecules, so that the average energy of motion of each molecule must be 7.5×10^{-8} ergs, and this represents an *average temperature,* for the sun's interior, of 375 million degrees. In 1907 Emden, by a different calculation, found that if the sun were made of air, the *temperature at its centre* would be 455 million degrees. Apart from details, it is clear that the interior temperature of a sun made of air would be one of hundreds of millions of degrees.

Yet there must be something wrong, for a simple calculation, of the type explained on p. 133, shews that the quanta of radiation which fly about at such temperatures would be energetic enough not merely to break up the molecules of air into atoms, but also to strip all, or nearly all, of the electrons from the atoms. At such temperatures each molecule of air would break up into its constituent nuclei and electrons just as surely as, on a hot day, a lump of ice breaks up into its constituent molecules. The electric forces which, in quieter surroundings, would unite the electrons and nuclei, first into atoms and then into complete molecules, find themselves powerless against the incessant hail of rapidly moving projectiles and the shattering blows of quanta of high energy; it would be like trying to build a house of cards in a hurricane. A sun consisting of molecules of air proves to be an inconsistency, a contradiction; our hypothesis has defeated itself, and we must start again from the beginning.

We may start wherever we like, but the conclusion which we must finally reach is that, no matter what kind of molecules the sun consists of, the heat at the sun's centre will break them up, either completely or nearly so, into their constituent nuclei and electrons. The same is true for all other stars, and this introduces an extreme simplification into the problem of the interior constitution of the stars. We cannot say how many complete molecules there are to a gramme without knowing the nature of the molecules, but once let these molecules be broken up into their constituent parts, and we know at once the total number of constituent parts, nuclei and electrons, which go to make up a gramme: it is about 3×10^{23}, regardless of the kind of molecule from which they originate.* Thus when the heat of a star has broken up its molecules into their constituent parts, we know the total number of such parts in the star, and it becomes easy to calculate the temperature of the star's interior, either from the theorem of Poincaré just mentioned or otherwise. The temperature will be the same as though the star were made of unbroken molecules of hydrogen.

Emden calculated in 1907 that the central temperature of a sun of this kind would be about 31,500,000 degrees. Later and more refined calculations by Eddington led to an almost identical temperature, but some still later calculations of my own give the substantially higher figure of 55,000,000

* This follows at once from the circumstance that the atomic weights of all elements are nearly double the atomic numbers. The statement is not true for hydrogen, but we can disregard the possibility of a star consisting of hydrogen.

degrees. There is no need for the moment to discuss which of these figures is nearest the truth. Their diversity will indicate what kind of degree of uncertainty attaches to all calculations of this type.

It is easy to see the physical necessity for this high temperature. The heat which flows away from the sun's surface must first have been brought there from its interior. Heat only flows from a hotter to a cooler place, and a vigorous flow of heat is evidence of a steep temperature gradient. The temperature must rise sharply as we pass from the sun's surface towards its centre, and this rise, continued along the whole 433,000 miles to the centre, must result in a very high temperature indeed being reached there.

The calculated central temperature of 30 to 60 million degrees so far transcends our experience that it is difficult to realise what it means. Let us, in imagination, keep a cubic millimetre of ordinary matter—a piece the size of an ordinary pin-head—at a temperature of 50,000,000 degrees, the approximate temperature at the centre of the sun. Incredible though it may seem, merely to maintain this pin-head of matter at such a temperature—i.e. to replenish the energy it loses by radiation from its six faces—will need all the energy generated by an engine of three thousand million million horse-power; the pin-head of matter would emit enough heat to kill anyone who ventured within a thousand miles of it.

High though this temperature is, calculations shew that it would not suffice to break up the stellar molecules completely. It would strip the atoms off all the electrons down to their *K*-rings (p. 126), but these would remain intact. It needs temperatures even higher than those we are now considering to strip the *K*-ring electrons from the nucleus of an atom. This result is true for the whole range, from about 30 to 60 million degrees, within which the temperature of the sun's centre is at all likely to lie, and it is true almost independently of the atomic weight or atomic number of the atoms of which we suppose the sun is to be built.

Thus if the sun is wholly gaseous, its central parts must consist of a collection of atoms stripped down to their K-rings, but not beyond, flying about independently like the molecules of a gas, and with them, also flying about like the molecules of a gas, all the stripped-off electrons which originally formed the L-ring, the M-ring, etc., of the atoms, the whole being at a temperature of somewhere between 30 and 60 million degrees. As we pass outwards towards the sun's surface we come to lower temperatures, at which the atoms are less completely broken up. Finally, close to the sun's surface, we may meet atoms which are completely formed except perhaps for one or two of their outermost electrons. In the surfaces of the coolest stars of all, we even find complete molecules, as, for example, the molecules of titanium oxide and magnesium hydride which shew themselves in the spectra of the red stars.

When the internal constitution of other stars is investigated in the same way, all main-sequence stars are found to have about the same central temperatures as the sun. Moreover, this is not the only property which they have in common. Fig. 23, which exhibits Seares' calculations of mean stellar densities, shews that the mean densities of main-sequence stars are all approximately the same, except for comparatively small deviations at the two extremities.

The mean density of the sun is 1.4, which means that the average cubic metre in the sun contains 1.4 tons of matter. At the sun's centre, the density is about 100 times this, so that a cubic metre there contains about 140 tons of matter.

For comparison, a cubic metre of lead contains only about 11 tons. If all stars were built on the same model as the sun, any two stars which had the same mean density would also have equal densities at their centres. But in stars having several times the weight of the sun, a new factor comes into play, namely, pressure of radiation—the pressure which radiation exerts in virtue of the weight it carries about with it. In most stars this pressure is insignificant in comparison with the pressure produced by the impact of material atoms and electrons, but in very massive stars it is large enough to influence the structure of the star. It is to this that the very massive stars whose diameters were tabulated on p. 256 owe their abnormally large size. It is a general consequence of the disturbing effects of radiation-pressure, that the weight of a very massive star is far more concentrated in its central regions than that of a lighter star, so that if a light and a massive star have the same average density, the latter will have by far the higher density at its centre. When this disturbing factor is allowed for, all stars in the upper part of the main-sequence are found to have approximately the same densities in their central regions, a density about equal to that at the centre of the sun, which we may estimate at 140 tons to the cubic metre. And we have already seen that the central regions of these stars have also approximately the same temperatures as the centre of the sun, whence it follows that their physical conditions are all substantially the same. Thus, the atoms in the central regions of all these stars must be broken down to the same extent as the atoms in the central regions of the sun. The K-rings of electrons survive intact, but the outer rings are transformed into a hail of electrons flying about like independent molecules.

With sufficient accuracy for our present purpose, all the

stars on the main-sequence, except perhaps those at its extreme lower end, may be supposed to be in the same physical condition. On account of this property, the main-sequence forms an admirable base-line from which to carry out a survey of the Russell diagram in respect of the physical conditions of stellar interiors.

Fig. 22 shews that a star to the right of the main-sequence has a greater diameter than a main-sequence star of the same weight. Consequently the energy it would emit in shrinking to its present diameter is less, and hence its molecular energy of motion is less (by Poincaré's theorem). It follows that its internal temperatures are lower, and its atoms are less completely broken up. Red giants such as Antares are found only to have central temperatures of from one to five million degrees, and their atoms probably retain intact not only their K-rings of electrons, but also their L-rings and part at least of their M-rings.

To the left of the main-sequence we come to a region in which stars, if they occurred at all, would have shrunk further, and so would have higher temperatures and more thoroughly broken atoms. Actually no stars are encountered until we come to the white dwarfs. Calculation shews that the central temperatures of these must be many hundreds of millions of degrees at least, and that their atoms must be stripped of electrons right down to the nuclei. Except for a small number of atoms which may have escaped this general fate, the stellar matter must consist of nuclei stripped absolutely bare, and of free electrons, all flying independently through the star. The high densities of these stars provide a convincing proof of the accuracy of this result. The mean density of Sirius B is certainly over 50,000, while that of van Maanen's star is probably over 300,000. There is no way in which matter can be packed as closely as this, except that of stripping the atoms of electrons right down to their bare nuclei.

The clearest general impression we can form of the Russell diagram in terms of physical condition is probably obtained as follows:

We think first of two detached bands of stars, one, the white-dwarf group, formed by stars in which all the electrons are torn off the atoms; and the other, the main-sequence, formed of stars in which the atoms are still surrounded by the K-rings of electrons, while the exterior rings have been torn off. Starting from about the middle of the main-sequence is the spur branch leading up to the red giants, as shewn in fig. 22. As we pass along this, the internal temperatures of the stars decrease, so that the stellar atoms are less broken up than in main-sequence stars. In the red giants at the extreme end, even M-ring electrons may still remain.

Russell's Hypothesis

Two entirely different explanations of this distribution of stars have been suggested. In 1925 Russell put forward a theory which centred primarily around the fact that the temperatures at the centres of the main-sequence stars are all very nearly equal. Let us simplify the situation for a moment by imagining it to be an ascertained fact that the temperatures at the centres of *all* stars are precisely the same, say 32,000,000 degrees. If this were a sure fact, it would be natural to conjecture that the stars had some sort of controlling mechanism by which they continually adjusted their central temperatures to this exact figure, so that if ever the temperature fell below 32,000,000 degrees the mechanism would come into play and raise the temperature to precisely this amount, while if it increased to above this figure, the mechanism would come into play and depress it. Such controlling mechanisms are of course common in engineering practice; there are for instance the safety-valves which keep the pressure in a boiler always uniform, the Watts-governor which keeps an engine going always at the same rate of speed, and the thermostat which keeps the temperature of a room constant.

A mechanism is already known for raising the temperature at a star's centre. If a star is not generating any energy at all in its interior, either by the annihilation of matter or otherwise, its emission of radiation causes it to shrink, and this, as we have seen (p. 270), causes its temperature to rise. Thus it is easy to keep a star's central temperature up to 32,000,000 degrees by arranging that no energy shall be generated so long as the temperature at the centre is below 32,000,000 degrees, and this is the main hypothesis on which Russell's theory is based. He supposes that no energy at all is generated by matter at temperatures below 32,000,000 degrees, but that, as soon as this temperature is reached, matter begins to annihilate itself in sufficient quantity to provide for the radiation of a star.

The trouble with the theory is that it seems impossible to regulate the temperature from the other end. A star whose central temperature is below 32,000,000 degrees must be contracting without generating heat. The contraction will not stop dead the moment the critical temperature is attained; its momentum will carry it on until the central temperature substantially exceeds 32,000,000 degrees. As soon as the temperature seriously exceeds 32,000,000 degrees at the centre, that of a substantial piece of the star will be

32,000,000 degrees or higher. The annihilation of all this matter must produce a profusion of heat which would raise the temperature of the star still further, resulting in more and more annihilation of matter, until finally the whole star disappeared in a flash of radiation. Indeed Russell's theory supposes that matter at 32,000,000 degrees is in a similar condition to gunpowder at its flash point. Mathematical analysis then shews that a star whose centre is at a temperature of 32,000,000 degrees would be in the state of a keg of gunpowder with a spark at its centre, and—well, "ohne hast, ohne rast" hardly describes the subsequent course of events.

Eddington has suggested that the stability of the stars might be saved by imagining a time-lag between the instant at which matter attained the critical temperature necessary for annihilation and the instant at which this annihilation occurred. It has not yet been proved that the proposed remedy could be made effective, but even if it could, other difficulties remain. As the normal star inhabits the main-sequence, Russell supposed it to be a property of normal matter to annihilate itself at a temperature of about 32,000,000 degrees, the supposedly uniform central temperature of all main-sequence stars. It then became necessary to introduce special *ad hoc* assumptions to explain the luminosity of white dwarfs and of stars on the red giant spur line, whose centres are at temperatures very different from 32,000,000 degrees. He accordingly supposed that such stars contained other types of matter which dissolved into radiation at temperatures which were respectively higher and lower than 32,000,000 degrees. Even if the stability difficulty could be overcome, this latter series of assumptions seems to me to be so artificial as to compel the abandonment of this interesting theory.

STARS

A discussion of the difficulties of Russell's theory led me to undertake a mathematical investigation of the stability of stars in general, and this was found to provide a simple and somewhat unexpected explanation of the otherwise incomprehensible distribution of stars in the Russell diagram; it is in brief that the unoccupied regions of the diagram represent stars in an unstable condition. I do not know what proportion of astronomers accept this explanation; some, whose opinion I value, do not. I do not think that much so far written in this book would be seriously challenged by competent critics, but it is only fair to say that at this point we are entering controversial ground.

The Hypothesis of Liquid Stars

Let us begin by imagining an enormous number of stars built on all possible plans, out of all kinds of substances. Mathematical investigation shews that some of these stars may be unable to shine with a steady light for either or both of two reasons—they may explode, like a heated keg of gunpowder, or they may have an inherent tendency to contract or expand without limit. Whether a star escapes the first pitfall or not depends mainly upon the properties of the substance of which it is built; whether it escapes the second depends mainly upon the way it is built. The two pitfalls are not altogether distinct, and when we consider the stability of wholly gaseous stars of enormously great weight, we find that the pits on the two sides of the path almost merge into one—only a narrow strip of safe ground is left between them. Nevertheless stars of enormously great weight are known to exist, and continue shining steadily. If then, these stars are wholly gaseous, they must occupy the one safe spot of ground between the two pits, and this informs us both as to the way they are built and as to the properties of the substance of which they are built.

We find that such stars only escape both pitfalls if their substance possesses properties which appear very improbable, and contrary to anything of which we have any experience or knowledge in physics; in brief, for such a star to remain stable, the annihilation of its matter must proceed at a rate which depends on the temperature. Such a property seems in every way contrary to the physical principles explained in Chapter II, as it is to all our expectations of atomic behaviour. The annihilation of matter is a far more violent change, and involves quanta of far higher energy, than mere radio-active disintegration, and as the latter process is not affected by temperature changes, it hardly seems possible that the process of annihilation should be, at any rate until we reach temperatures of the order of the 2,200,000,000,000 degrees tabulated on p. 135.

We have, however, already found indications that the stars are not purely gaseous, since purely gaseous masses could not form close binary systems of the type observed in the spectroscopic binaries (p. 211). Such systems can only be formed out of a mass which simulates the properties of a liquid rather than those of a gas; the mass need not be wholly in a liquid state, but there must be a considerable divergence from the state of a pure gas, at any rate in its central regions. Additional evidence to the same effect will also emerge later (pp. 293, 294).

As soon as we admit that the interiors of the stars need not be in a completely gaseous state, the whole situation changes, even a slight departure from the gaseous state being

---

* This provides a further objection to Russell's hypothesis, which, to avoid confusion, was not mentioned on p. 279.

found to impart a great deal of additional stability to the star. If a star of great weight is purely gaseous in its structure, the region of stability between the two pitfalls is reduced to a narrow strip, and only by treading this can the star escape the alternative fates of exploding or collapsing. But if the star has a liquid, or partially liquid, centre, this strip of safe land is so wide that, consistently with stability, the stellar material may have exactly the property that we should *à priori* expect to find, namely, that its annihilation proceeds, like radio-active disintegration, at the same rate at all temperatures. If the substance of the star has this property, the star can be in no danger of exploding, for a mass of uranium or radium does not explode whatever we do to it. And mathematical analysis shews that if the centre of the star is either liquid, or partially so, there is no danger of collapse; the liquid centre provides so firm a basis for the star as to render a collapse impossible.

These considerations suggest the two complementary hypotheses:

1. That the annihilation of stellar matter proceeds spontaneously, not being affected by the temperature of the star.

2. That the central regions of the stars are not in a purely gaseous state; their atoms, nuclei and electrons are so closely packed that they cannot move freely past one another, as in a gas, but rather jostle one another about like the molecules of a liquid.

If we have been right (p. 138) in attributing the observed highly penetrating radiation in the earth's atmosphere to the annihilation of matter in distant astronomical bodies, then the first hypothesis is confirmed. For the radiation could not retain its observed high penetrating power if it had already penetrated any great thickness of matter. The struggle of passing through matter lengthens the wave-length of all kinds of radiation (the quanta get weaker all the time), and so diminishes its penetrating power. Thus wherever the highly penetrating radiation originated, it must have got out into empty space without much of a struggle, and this is the same thing as saying that it must have originated in matter at a comparatively low temperature. Thus the existence of the highly penetrating radiation proves that matter can be annihilated in great quantities at quite low temperatures; the high temperatures of stellar interiors are not needed, as Russell's theory asserts, for annihilation to occur.

A simple calculation shews that there can be no appreciable annihilation of the earth's substance. In the sun about one atom in every 10^{19} is annihilated every minute; if even a ten-thousandth part as many atoms as this were annihilated in the earth, its surface would be too hot for human habitation. We can no longer explain this by saying that the sun is hot and the earth cool, so that annihilation goes on in the former but not in the latter. We must rather suppose that the atoms in the sun are of a different type from those on earth. Solar atoms spontaneously annihilate themselves, terrestrial atoms do not, or at least do not to any appreciable extent.

The Stability of Stellar Structures

For the present, let us tentatively accept the hypothesis that the generation of stellar energy occurs spontaneously, like the disintegration of radio-active atoms. The atoms which are responsible for the light and heat of the stars may be regarded as super-radio-active atoms which spontaneously annihilate themselves and so change their substance into radiation.

We have already seen that, on this view of the mechanism of generation of stellar energy, a star can only continue to shine steadily if its central regions are not in a purely gaseous condition. A star built on foundations of highly compressible gas meets the same fate as a house built on sand; it collapses. A purely gaseous star is a dynamically unstable structure, and must continually contract until the atoms in its central regions are so closely packed that their state can no longer be regarded as gaseous. Then, and then only, can the star exist permanently as a stable structure. Thus the central regions of any actual permanent star, the sun for instance, must be in a state which for brevity we may describe as liquid.

Now let us imagine the sun to be expanded to ten times its present diameter. This would diminish its density to a thousandth part of its original value. The actual sun is 40 per cent. more dense than water, but the expanded sun would only be as dense as ordinary atmospheric air. The atoms and electrons, having moved ten times farther apart, would be so distant from one another that the new sun might be regarded as wholly gaseous. Thus it would be dynamically unstable and could not remain in its wholly gaseous state.

Our imaginary expanded sun is of course no longer a main-sequence star in the Russell diagram. In expanding the sun to ten times its present size we move it off the main-sequence into a region entirely vacant of stars—in fact, into the great gulf which lies between the red giants and the red dwarfs (see fig. 22, p. 261). Thus, it appears that even if we deliberately place a star in this region, it does not stay there but immediately contracts until it gets on to the main-sequence. May not this explain why the region in question is untenanted by stars?

Next let us imagine the sun contracted to a tenth of its present diameter, so that its atoms and electrons move ten times nearer to one another. Its mean density is thereby increased from 1.4 to 1400 times that of water, and its central density from about 140 to 140,000. You may check me here by pointing out that if the sun is already in a liquid state it cannot be compressed to any such extent—a liquid cannot usually have its density increased a thousand-fold. But we have already noticed that halving a star's diameter doubles its temperature throughout. In the same way reducing a star's diameter to a tenth increases its temperature ten-fold, so that the sun's central temperature will be increased from, say 50 million degrees to 500 million degrees. And at this latter temperature atoms hardly exist any longer as such—the stellar matter consists almost entirely of free electrons and nuclei. And these are so minute, that the increase of the sun's mean density from 1.4 to 1400 times the density of water is not only possible, but leaves the sun's substance in a state which may best be described as gaseous. Once again, then, the new sun is dynamically unstable. It would be represented by a point well to the left of the main-sequence, near the middle of the unoccupied region between the main-sequence and the white dwarfs, but as it is unstable it cannot maintain its position here. Again we see that even if we place a star in this region it cannot stay there. And, again—may it not be that the reason why this region is unoccupied is that it represents unstable stars?

Once more you may check me. If I have made my point, it has been by the help of the rise of temperature which accompanies contraction. When we imagined the sun to expand, ought we not to have considered the fall of temperature which accompanies expansion? The answer is that we ought, but it would have made no difference. Lowering the temperature will cause a number of L-rings, and possibly also of M-rings, of electrons to re-form, so that the new atoms will be of larger size, but they will not lose their freedom of motion sufficiently to make the sun stable. It would have been different if we had been discussing a star of 10 or 50 times the sun's weight; then it can be shewn that the re-formation of K- and L-rings would have produced a series of stable configurations. And the spur branch in the Russell diagram exists to provide a home for just such stars.

The whole problem is too complicated to be discussed satisfactorily in this fragmentary way; its proper discussion involves very complicated mathematical analysis. Mathematical discussion shews that the Russell diagram can be divided into regions representing stable and unstable configurations in the manner shewn in fig. 24.

The unstable areas are so marked; the remaining areas are stable. The dots which form a sort of background to the diagram represent 2100 stars whose absolute magnitudes are known through their parallaxes having been determined spectroscopically at Mount Wilson. The observational material is not perfect, for considerable uncertainty attaches to all spectroscopic parallaxes of B-type stars, and A-type stars are almost unrepresented because it is practically impossible to obtain their parallaxes by the spectroscopic method. The theoretical curves are probably still more imperfect, yet, such as they are, they seem to suggest very forcibly that the occupied and unoccupied regions coincide with those representing stable and unstable configurations; after making all possible allowances for the imperfections both of theory and of observation, too much agreement remains to be explained away as mere coincidence.

Thus the conclusion to which mathematical discussion seems to lead is that the regions in the Russell diagram which are occupied, represent stars whose central regions are in a

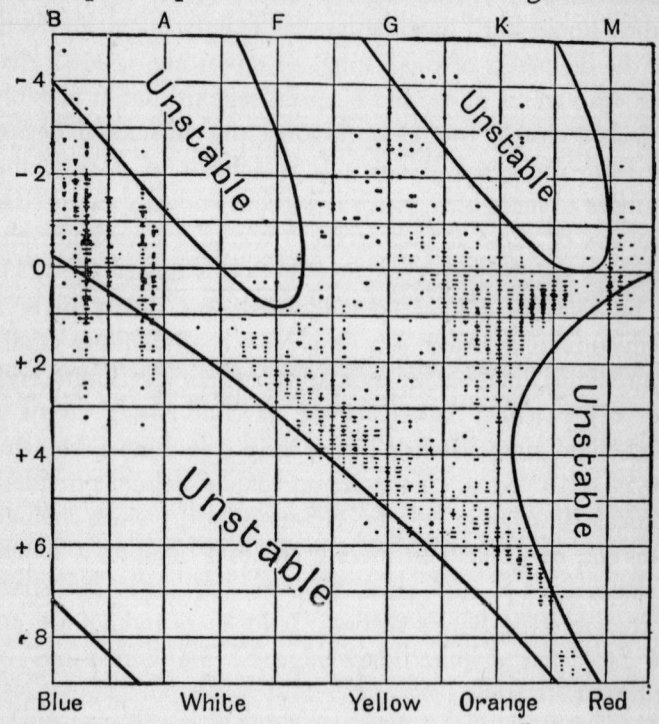

FIG. 24.—Stable and unstable configurations in the Russell diagram.

liquid, or nearly liquid, state. All other stars are unstable, so that the corresponding regions in the Russell diagram are necessarily vacant. To put it in less technical language, all the stars in the sky must have liquid, or nearly liquid, centres.

Here we have a piece of the puzzle which seems to fit on to

the piece we unearthed in Chapter IV, where we found that a star could only break up by fission if it had a liquid, or nearly liquid centre. Evidence accumulates that the stars have liquid rather than gaseous centres.

Criticism of the foregoing hypothesis—generally described as the "liquid-star" hypothesis—has mainly taken the form that the diameters of the K-rings of atoms are so small that the K-ring atoms in the sun's central regions cannot possibly be packed closely enough to involve any substantial departure from the gaseous state. It is difficult to discuss, and still more to meet, this criticism without knowing the precise diameters of these K-ring atoms. We of course know the diameters assigned to the K-ring by Bohr's theory (p. 121), but no one any longer contends that this theory gives a true picture of the atom. It provides a good working model within limits, but we do not know where the limits end. The only practical experience we have of K-ring atoms is with helium atoms; Bohr's theory assigns to these a diameter of 0.54×10^{-8} cms. Yet solid and liquid helium provide a practical illustration of the closeness with which helium atoms can be packed; in these each atom occupies a sphere of diameter 4×10^{-8} cms., or over 400 times the space allotted to it by Bohr's theory. It looks as though we are still far from definite knowledge of the dimensions of K-rings of electrons.

The new wave-mechanics of Schrödinger draws a very different picture of atomic interiors from the simpler theory of Bohr which it is rapidly superseding. Even the electron is something very different from the electron of Bohr's old theory. It is the old-fashioned electron only when it is at an infinite distance from the nucleus. As it gradually approaches this nucleus, it undergoes a metamorphosis of a kind which no one has yet succeeded in describing, and it is utterly impossible to say what form it may have assumed by the time it is doing what we call "describing a K-ring orbit." All we know about the K-ring orbit is its energy, and it seems impossible to predict the amount of space occupied by such an orbit until we have a better knowledge of the qualities of the article which is describing it.

We have of course to admit that the physical evidence, such as it is, seems to point to K-ring atoms being substantially smaller than is needed for the liquid-star hypothesis. But the astronomical evidence seems to me stronger and more reliable, and to point in exactly the opposite direction. And here we must leave the puzzle until further pieces come to light.

Until we know the kind of atoms of which a particular star is composed, we cannot calculate the extent to which they will be broken up by the temperature of the star's interior. As a consequence, the theoretical curves of demarcation between stable and unstable configurations cannot be calculated without assuming definite atomic numbers for the stellar atoms.

The curves shewn in fig. 24 have been drawn for an atomic number of about 95, this being slightly higher than the atomic number, 92, of uranium. This atomic number was selected because it was found to produce the best agreement between theory and observation, but we shall see that other considerations justify our choice.

Stellar Structure

A star, like a house or a pile of sand, is a structure which would collapse under it own weight were it not that each layer is held up against gravity by the pressure which the next inner layer of the star exerts upon it. This pressure is not, like ordinary gas-pressure, the result of the impacts of complete molecules. It is produced in part by the impact of a certain number of atoms which have been stripped of electrons almost or quite down to their nuclei, but to a far greater extent by the impact of a hail of free electrons. In massive stars, an additional pressure is produced by the impact of radiation which, as we have seen, carries weight about with it, and so exerts pressure on any obstacle it encounters. The combined impacts of free electrons, of atoms (or bare nuclei), and of radiation prevent the star from falling in under its own gravitational attraction.

This gives a reasonably good snapshot picture of a star's structure. The corresponding picture of its mechanism is obtained by thinking of the nuclei as α-ray particles, of the free electrons as β-ray particles, and of the radiation as γ-rays (although in most stars the main bulk of the radiation has the wave-length of X-rays). All these thread their way through the star, and, precisely as in laboratory work, the γ-rays are more penetrating than the β-rays, and the α-rays are more penetrating than either.

The Transport of Heat in a Star. We have seen how the heat of a gas is merely the energy of its molecular motion. Conduction of heat in a gas is usually studied by regarding each molecule as a carrier of energy; when it collides with a second molecule the energy of the two colliding molecules is redistributed between them, and in this way heat is transported from hotter to cooler regions. Each molecule has a power of transport which is jointly proportional to its energy of motion, its speed of motion, and its "free-path"—the distance it travels between successive collisions.

In the interior of a star, there are three distinct types of carrier in action—atoms (or bare nuclei), free electrons, and radiation. We can compare their relative capacities as carriers by multiplying up the energy, speeds and free-paths of each. For this purpose the "free-path" of radiation may be taken to be the distance the radiation travels before 37 per cent. of it has been absorbed, since it can be shewn that this is the average distance it carries its energy. On carrying out the calculation, the carrying capacity of both nuclei and electrons is found to be insignificant in comparison with that of the radiation. The nuclei and electrons may have the greater amount of energy to carry, but owing to their feebler penetrating powers, the distance over which they carry it, their free-path, is far less than that of the radiation. Their speed of transport is also less, since radiation transports its energy with the velocity of light. In this way it comes about that practically the whole transport of energy from the interior of a star to its surface is by the vehicle of radiation.

This general principle was first clearly stated by Sampson in 1894. He also shewed how the temperature of any small fragment of a star's interior must be determined by the condition that it receives just as much radiation as it emits, but his detailed applications were vitiated through his using an erroneous law of radiation. Twelve years later, Schwarzschild independently advanced the same idea, and expressed it in mathematical equations of "radiative equilibrium" which have formed the basis of every subsequent discussion of the problem.

Just because radiation completely outstrips atoms and electrons in carrying energy from a star's interior to its surface, it follows that the build of a star must be determined by the opacity of the matter in its interior. If this is altered, the carrying power of the radiation is altered, and this affects the whole structure of the star. A star whose interior was entirely transparent could not retain any heat at all; its whole interior would be at a very low temperature and the star would be of enormous extent. On the other hand, in a very opaque star, all energy would remain accumulated at the spot at which it was generated, so that the interior temperature would become very high and the star's diameter would be correspondingly small. It is, of course, the intermediate cases which are of practical interest, but the extreme instances just mentioned shew how a star's build depends on its opacity.

Unfortunately it is impossible to obtain any sort of direct measurement of the opacity of stellar matter. We cannot even measure the opacity of terrestrial matter under stellar conditions, since the interiors of the stars are at incomparably higher temperatures than any available in the laboratory. However, we know that the opacity of stellar matter is due to the atoms, nuclei and free electrons of which it is composed checking the onward journey of radiation, and although we cannot obtain a sample of stellar matter, we know fairly definitely how many atoms, nuclei and electrons such a sample would contain. Thus it becomes a matter of theoretical calculation to determine its opacity.

Such a calculation was carried through by Dr. Kramers of Copenhagen in 1923, and his results have gained general acceptance. In so far as they can be tested in the laboratory, they agree well with observation. And, although there is a big gap between laboratory conditions and stellar conditions, it is difficult to see how Kramers' formula could fail in the stars.

From this formula we can determine the build of the stars completely, or, if the build of the star is supposed to be known, Kramers' formula tells us the rate at which energy flows to its surface (this depending entirely on the opacity of the star's substance), and this in turn tells us at what rate energy must be generated inside the star for it to be able to remain in equilibrium in the configuration in question. As might be expected, configurations of different diameters are found to require different rates of generation of energy. In nature, a star must adjust its diameter to suit the rate at which it is generating energy; in so doing it fixes not only its diameter but also its surface-temperature, colour and spectral type. If a star's rate of generation of energy were suddenly to change, the star would expand or contract until it had assumed the radius and temperature suited to its new rate of generation of energy.

Detailed calculation shews that for wholly gaseous stars, large diameters correspond to feeble generation of energy, and *vice versa*. Thus if the stars were wholly gaseous, red giants would be less luminous than main-sequence stars of the same weight. Seares' diagram reproduced on p. 266 shews that the reverse is actually the case, a red giant emitting from 10 to 20 times as much total radiation as an equally massive main-sequence star. This provides evidence against the stars being wholly gaseous, but there is stronger evidence than this. For wholly gaseous stars, the thick lines shewn in Seares' diagram would be straight slant lines, slanting upwards to the left. The wide divergence between such a system of slant lines and the curves shewn in fig. 23 gives some indication of the extent to which the condition of stellar matter diverges from the purely gaseous state.

According to Kramers' theory, the opacity of matter depends on the atomic numbers and atomic weights of the atoms of which it is built, a large clot of matter in the form of a massive atomic nucleus being far more effective in absorbing radiation than a large number of small clots of the same total weight. Everyday terrestrial experience shews that this is so. It is for this reason that the physicist and surgeon both select lead as the material with which to screen their X-ray apparatus; they find that a ton of lead is far more effective in stopping unwanted X-rays than a ton of wood or of iron. If we knew the strength of an X-ray apparatus, and the total weight of shielding material round it, we could form a very fair estimate of the atomic weight of the shielding material by measuring the amount of X-radiation which escaped through it.

A very similar method may be used to determine the atomic weights of the atoms of which the stars are composed. A star is in effect nothing but a huge X-ray apparatus. We know the weights of many of the stars, and the rate at which they are generating X-rays is merely the rate at which they are radiating energy away into space. If we could cut each atomic nucleus in a star into two halves, we should halve the opacity of the star, so that radiation would travel twice as far through the star before being absorbed. If the star were wholly gaseous, this would result in its expanding to four times its original diameter, and in its surface-temperature being halved. It follows that we can calculate the atomic weight of the atoms of which a star is composed from the weight, luminosity and surface-temperature of the star.

The atomic weights of a number of stars, which I calculated on the supposition that the stars were wholly gaseous, came out in practically every case higher than that of uranium, which is the weightiest atom known on earth. They not only proved to be higher, but enormously higher; so high indeed, as to seem utterly improbable. Again the explanation seems to be that the stars are not wholly gaseous. As soon as stellar interiors are supposed to be partially liquid, the calculated atomic weights are reduced enormously. They can no longer be determined exactly, but the atomic number of about 95 to which we were led from a consideration of the Russell diagram seems to be entirely consistent with all the known facts.

Indeed other considerations seem to suggest that the atomic numbers of stellar atoms must be higher than 92. *A priori* stellar radiation might either originate in types of matter known to us on earth or else in other and unknown types. When once it is accepted that high temperature and density can do nothing to accelerate the generation of radiation by ordinary matter, it becomes clear that stellar radiation cannot originate in types of matter known to us on earth. Other types of matter must exist, and, as, with two exceptions, all atomic numbers up to 92 (uranium) are already occupied by terrestrial elements, it seems probable that these other types must be elements of higher atomic weight than uranium.

These super-heavy atoms must not be expected to disclose their presence in stellar spectra, for these only inform us as to the constitution of the atmospheres of the stars. And as the lighter atoms float to the top it is these, in the main, which figure in stellar spectra. If the sun's atmosphere had contained any considerable number of super-heavy atoms when the planets were born, some of them ought still to exist in the earth. There cannot be any great number, or their high generation of energy would betray them. The simplest view seems to be that the heavier atoms sink to the centre in the stars, and that the earth was formed mainly or solely out of the lighter atoms which had floated to the sun's surface.

Stellar Evolution

We have supposed that the stars were born initially as condensations in the outer fringes of spiral nebulae. These condensations would, from the mode of their formation, necessarily be of all sorts of sizes, and subject to the single restriction that none of them could be below a certain limit of weight. Thus we should not expect the stars, either at birth or subsequently, to be all of the same size or weight, or all in the same physical condition. The stars would start their existences at different points in the Russell diagram, but we may imagine that their initial positions are limited to those parts of the diagram which can be occupied by stars—either, as the liquid-star hypothesis would suggest, because these are the only stable configurations, or else for some other reason so far undiscovered. Each year a star loses a certain amount of weight, and its rate of generating energy, and so also its luminosity, are correspondingly reduced, with the result that it moves to a new position in the diagram. Turning to Seares' diagram of stellar weights on p. 266, we may think of the curves of equal weight as a flight of steps—very uneven steps, it is true—each representing a lower weight than the one above. Whatever a star's evolution may be, it is essential that it should always be *down* the steps: any upward step is impossible.

We can trace out two possible paths of stellar evolution in the Russell diagram which involve no entry into regions unoccupied by stars—two roads along which the stellar army may march as they transform their substance into radiation. The first is of course the "main-sequence," which a great number of considerations suggest to be the main line of march of the stellar army. The branch which starts from the red giants in the Russell diagram represents a second possible line of march, a certain number of stars travelling along this branch until they reach the main-sequence as blue or white stars, and then travelling down the lower half of the main-sequence to end as faint red stars passing on to ultimate extinction.

Progress along each of these roads is accompanied by a continuous shrinkage in the size of the star, its diameter steadily decreasing. This is not the same thing as saying that the star's density continually increases, for the star is continually diminishing in weight, so that even if the star's density remained the same, its diameter would decrease. Nevertheless a study of Seares' determinations of mean densities, as shewn in the diagram on p. 266, suggests that there is a continuous increase of density, although this becomes very slight in the middle reaches of the main-sequence.

Practically every theory of stellar evolution which has ever been propounded has imagined the march of the stellar army to be of the same general type as that just described, although perhaps present-day opinion is inclined to treat the main-sequence as the principal line of march, whereas earlier theories supposed the youngest stars to march solely along the red-giant branch, only joining the main-sequence with middle age. The first serious theory of all, that of Lockyer, was expressed in terms of branches of ascending and descending temperature, these together forming the last-mentioned line of march in the Russell diagram. A theory which Russell propounded in 1913 again assigned to the stars the lines of march just described. It also attempted a physical explanation, since abandoned, as to why the stars followed these particular paths rather than others. His more recent theory of 1925 only differed from his earlier theory in giving the new explanation, which we have already discussed (p. 276), as to why the stars followed these particular paths.

At present, it is probably fair to say that nearly, and perhaps quite, all astronomers are agreed that the evolutionary paths of the stars are of the general type we have described. Some stars start as red giants, some as blue, some possibly in intermediate conditions. As they age, all move downwards in the Russell diagram, their various paths converging to a point at the fork of the reversed γ shewn in fig. 22, and after passing this point they move down the main-sequence. On the other hand, there is the widest difference of opinion as to the physical interpretation which is to be assigned to these paths. Most astronomers are probably suspending judgment until some definite observational evidence is obtained to decide between conflicting theories.

When the stars first came into being as flecks of fiery spray thrown off by spinning nebulae, they would consist of mixtures of atoms of all kinds, some perhaps being so short-lived as to transform themselves almost at once into radiation, and others having such long lives that they may properly be described as permanent. Except for a small number of radio-active atoms, the earth must consist entirely of atoms of this latter type. Calculation shews that terrestrial atoms must have enormously longer lives than the average stellar atom, otherwise their self-annihilation would make the earth too hot for habitation. The permanent atoms in a star contribute almost nothing to its energy-generating capacity, and so merely add to its weight. The shortest lived atoms of all contribute greatly to the star's generation of energy while adding but little to its weight. In general the shorter the life of any type of atom, the greater the proportion of its numbers annihilated per year, and so the greater the amount of energy it generates per ton of weight.

A star begins life with a large proportion of short-lived atoms, and so at first generates energy furiously. As it ages, the shortest lived atoms disappear first, and in so doing reduce the average energy-generation of the star per ton so that, as a star's weight decreases, so also must its rate of generation of energy per ton. Finally all the atoms with much energy-generating capacity have disappeared, and the star is left, a shrunken and diminished mass of atoms which have very little capacity for generating radiation.

To put the same thing in another way, the rate at which a star generates energy per ton is proportional to the death-rate in its population of atoms. To say that Sirius generates 16 times as much energy per ton as the sun is only another way of saying that the average atom in Sirius has only a sixteenth of the expectation of life of the solar atoms; their death-rate is 16 times as high. As those types of atoms which have the highest death-rate gradually die off in any star, the average death-rate of the population decreases, or, in other words, as a star ages, its capacity for energy-generation per ton decreases.

This agrees with the findings of observational astronomy.

STARS

The most massive stars not only generate more energy than less masssive stars, as is in any case to be expected; they also generate enormously more energy per ton. This is illustrated by the following list of main-sequence stars:

Star	Weight (in terms of sun)	Generation of energy (ergs per gramme)
Pearce's Star A	36.3	15,000
V Puppis A	19.2	1,000
Sirius A	2.45	29
Sun	1.00	1.90
ϵ Eridani	(0.45)	0.26
Kruger 60 B	0.20	0.021

Seares' diagram of stellar weights (p. 266) shews that this is a general property of the stars. To repeat our former metaphor, the stars squander their substance lavishly in their youth, while they have plenty left to spend, but parsimony comes over them with old age. Theoretical considerations have now given us an explanation of this phenomenon.

The same diagram shews that two stars of the same weight do not usually have the same luminosity. In general, giant stars on the spur branch leading out to the red giants have substantially higher luminosities than the main-sequence stars of equal weight. We have already noticed how a red giant may emit as much as 10 or 20 times the radiation of an equally massive main-sequence star. The same story is repeated when we pass from the main-sequence stars to the white dwarfs. Main-sequence stars emit enormously more radiation—anything up to 500 times more—than white dwarfs of equal weight. This is illustrated by the three following white dwarfs, which may be compared with the last three stars of the preceding table:

Star	Weight (in terms of sun)	Generation of energy (ergs per gramme)
Sirius B	0.85	0.0027
o$_2$ Eridani B	0.44	0.002
van Maanen's star	(0.20)	(0.00055)

We have hitherto supposed generation of energy to be spontaneous and so unaffected by changes of physical conditions. Yet the facts just mentioned seem to suggest that this can hardly be the whole truth of the matter. To state the objection in terms of a concrete instance, Sirius A and its white dwarf companion Sirius B must in all probability have been born at the same time out of the same nebula (p. 268), yet the former generates 4000 times as much energy per ton as the latter. It seems improbable that so great a difference can be attributed to different types of atoms; the common origin of the two stars almost precludes this. We know that the atoms are in different physical conditions in the two stars; in Sirius A they have retained their K-rings intact, while in Sirius B, the white dwarf, they are completely broken up into bare nuclei and free electrons. If the two components of Sirius consist of essentially the same types of atoms, as their common origin would lead us to expect, then the enormous difference in the rates at which these atoms generate energy would seem to depend on the different physical conditions of their atoms.

The considerations brought forward in Chapter III make it highly probable that a star's rate of generation of energy depends on the physical condition of its atoms. We there supposed stellar energy to be generated through electrons coalescing with protons; protons exist only in atomic nuclei, and purely physical considerations led to the conjecture that the only electrons which can coalesce with a particular proton are those which are momentarily describing orbits around the nucleus in which the proton resides. A study of stellar structure supports this hypothesis, for if energy could be generated by free electrons falling into nuclei, it can be shewn that the whole star would be unstable and would explode in a flash of radiation. On this hypothesis, a star in which only a few atoms have any electrons left in orbital motion can of course generate but little energy. This at once explains the feeble energy-generating powers of the white dwarfs, and also gives an inkling as to why the red giants, in which L- and M-rings of electrons survive, generate more energy than main-sequence stars of equal weights.

As a star ages and its weight decreases, it continually has to pick out new configurations such as make its emission of energy equal to its internal rate of generation of energy. The same star may be a red giant, a main-sequence star and a white dwarf in turn. Stripped of technicalities this means that a star continually adjusts its diameter to suit its varying rate of generation of energy.

On the hypothesis just considered, a star alters both its emission and its generation of energy on changing its diameter. At every instant it has to select a diameter for which the two exactly balance. The star has so large a range of rates of generation, according as it has few or many electrons left in orbital motion, that it is likely always to be able to find a configuration of equilibrium. At any rate all the stars in the sky appear to have done so, with the exception of the long period variables which are continually expanding and contracting as though they could not hit upon a diameter at which their income and expenditure of energy would just balance.

This same hypothesis immediately makes it possible for all the great variety of stars in the galactic system to be of approximately the same age, and so to have been all born out of the same nebula. The most luminous galactic stars can hardly have been generating energy at their present rate for more than about 100,000 million years—any longer age would require an impossibly high weight to start with. Yet the motions of the stars indicate that even these highly luminous stars must have been in existence for at least 50 times this period. The apparent contradiction disappears if we admit that the extreme luminosity of the brightest stars may be a recent development, and that for perhaps 98 per cent. of its life such a star was losing but little energy because most of its atoms were stripped bare of electrons and so were immune from annihilation. The requisite proportion of 98 per cent. may seem suspiciously large, but we have stated an extreme case; we need only demand so large a proportion in the case of a very rare type of star: probably not more than one star in ten million is of such a type.

In their earlier dormant state, these stars, which are now so luminous, would in effect have been white dwarfs of enormous weight. Observational astronomy provides no evidence that any such stars exist, but there is certainly no evidence that they do not exist. Very massive stars are known to be very rare objects, so that in all probability we should have to travel a long distance from the sun before finding one, and then it might be so distant as to be invisible from the earth. In any case, a very distant star of feeble luminosity would be exceedingly likely to escape detection. The fact that none have so far been found does not prove that none exist.

Moreover, it is far from absolutely certain that such stars

have not been found. Very massive white dwarfs ought to have higher surface-temperatures than either massive main-sequence stars or than the known white dwarfs, all of which are of small weight. A whole group of stars is known—the O-type stars—whose spectra indicate very high temperatures indeed. These are usually interpreted as stars of enormous luminosity at enormous distances, but it is possible that some at least of them may be stars of feeble luminosity at moderate distances. In particular the central stars of the planetary nebulae are of types O and B and yet, according to all measurements, are less luminous than the sun, although the normal main-sequence star of these spectral types is generally about a thousand times as luminous as the sun. Thus there is a possibility that the planetary nebulae may be stars of the kind we need, although we cannot overlook two serious objections to such a view. The first is that, if we interpret their spectra in the ordinary way, they appear to be moving with enormous velocities which seem quite inappropriate to exceptionally massive stars; the second is that their spectra do not exhibit the general displacement to the red which, on the theory of relativity, ought to be shewn by stars of such great weight and small diameters. In spite of these difficulties, however, it still seems possible that either these, or other O- or B-type stars of feeble luminosity, may be very massive stars in the dormant condition contemplated by the hypothesis we have just been discussing. In brief, we imagine that a massive star may have its weight conserved through existing as a planetary nebula or a dwarf O-type star for millions of millions of years, and then burst out as a highly luminous star with all the appearance of extreme youth. But there is at present insufficient observational evidence either for or against such a hypothesis. Some pieces of the puzzle are missing, and we can only wait until they turn up.

White Dwarfs. Apart from these hypothetical massive white dwarfs, astronomers generally regard ordinary white dwarfs as the final stage in stellar evolution. There is general agreement that they are stars with central temperatures so high that their atoms are stripped bare of electrons, but there is no general consensus of opinion as to why stars shrink to this condition.

On the liquid star hypothesis, the unoccupied regions in the Russell diagram represent unstable configurations. Usually a slight loss of weight by a star merely moves it to a new position in the diagram contiguous to the old one. Sometimes, however, this slight move may happen to carry the star into an unstable region of the diagram, in which case it will hurriedly traverse this region, until finally it ends up in some entirely different stable configuration.

The liquid star hypothesis explains the white dwarf state quite simply as the final state to which a star shrinks cataclysmically when its generation of energy is no longer sufficient to entitle it to a place in the main-sequence. In this state the star radiates so little energy that annihilation and decay are almost entirely checked. We have seen that if the sun went on radiating at its present rate for 15 million million years, its whole weight would be transformed into radiation. By contrast, van Maanen's star can, and probably will, go on radiating at its present rate for 15 million million years without losing more than about a thousandth part of its present weight. We may think of the white dwarf state as a final state from which change and decay have so nearly disappeared that a star which shrinks to this state acquires a new lease of life for a period of thousands of millions of millions of years—we can only wonder to what purpose.

CHAPTER VI

Beginnings and Endings

WE have seen how the solid substance of the material universe is continually dissolving away into intangible radiation. The sun weighed 360,000 million tons more yesterday than to-day, the difference being the weight of 24-hours' emission of radiation which is now travelling through space, and, so far as direct observation goes, is destined to journey on through space until the end of time. The same transformation of material weight into radiation is in progress in all the stars, and to a lesser degree on earth, where complex atoms such as uranium are continually changing into the simpler atoms of lead and helium, and setting radiation free in the process. But against the sun's daily loss of weight of 360,000 million tons, the earth is only losing weight from this cause at the rate of about ninety pounds a day.

Cyclic Processes. It is natural to ask whether a study of the universe as a whole reveals these processes as part only of a closed cycle, so that the wastage which we see in progress in the sun and stars and on the earth is made good elsewhere. When we stand on the banks of a river and watch its current ever carrying water out to sea, we know that this water is in due course transformed into clouds and rain which replenish the river. Is the physical universe a similar cyclic system, or ought it rather to be compared to a stream which, having no source of replenishment, must cease flowing after it has spent itself?

Thermodynamics

To this question, the wide scientific principle known as the second law of thermodynamics provides an answer in very general terms. If we ask what is the underlying cause of all the varied animation we see around us in the world, the answer is in every case, energy—the chemical energy of the fuel which drives our ships, trains and cars, or of the food which keeps our bodies alive and is used in muscular effort, the mechanical energy of the earth's motion which is responsible for the alternations of day and night, of summer and winter, of high tide and low tide, the heat energy of the sun which makes our crops grow and provides us with wind and rain.

The first law of thermodynamics, which embodies the principle of "conservation of energy," teaches that energy is indestructible; it may change about from one form to another, but its total amount remains unaltered through all these changes, so that the total energy of the universe remains always the same. As the energy which is the cause of all the life of the universe is indestructible, it might be thought that this life could go on for ever undiminished in amount.

Availability of Energy. The second law of thermodynamics rules out any such possibility. Energy is indestructible as regards its amount, but it continually changes in form, and generally speaking there are upward and downward directions of change. It is the usual story—the downward journey is easy, while the upward is either hard or impossible. As a consequence, more energy passes in one direction than in the other. For instance, both light and heat are forms of energy, and a million ergs of light-energy can be transformed into a million ergs of heat with the utmost ease; let the light fall on any cool, black surface, and the thing is done. But the reverse transformation is impossible; a million ergs which have once assumed the form of heat, can never again assume the form of a million ergs of light. This is a special example of the general principle that radiative energy tends always to change into a form of longer wave-length, never into a form of shorter wave-length. When blue light falls on a fluorescent substance, it emerges as green, yellow or red light, but the reverse transition is unknown; fluorescence always increases the wave-length of the light (Stokes' law). We have seen how fluorescent substances become visible when they are placed in the ultra-violet region of a spectrum, but no known substance becomes visible when placed in the infra-red.

It may be objected that the everyday act of lighting a fire disproves all this. Has not the sun's heat been stored up in the coal we burn, and cannot we produce light by burning coal? The answer is that the sun's radiation is a mixture of both light and heat, and indeed of radiation of all wave-lengths. What is stored up in the coal is primarily the sun's light and other radiation of still shorter wave-length. When we burn coal we get some light, but not as much as the sun originally put into the coal; we also get some heat, and this is more than the amount of heat which was originally put in. On balance, the net result of the whole transaction is that a certain amount of light has been transformed into a certain amount of heat.

All this shews that we must learn to think of energy, not only in terms of quantity, but also in terms of quality. Its total quantity remains always the same; this is the first law of thermodynamics. But its quality varies, and tends to vary always in the same direction. Turnstiles are set up between the different qualities of energy; the passage is easy in one direction, impossible in the other. A human crowd may contrive to find a way round without jumping over turnstiles, but in nature there is no way round; this is the second law of thermodynamics. Energy flows always in the same direction, as surely as water flows downhill.

Part of the downward path consists, as we have seen, of the transition from radiation of short wave-length into radiation of longer wave-length. In terms of quanta (p. 119) the transition is from a few quanta of high energy to a large number of quanta of low energy, the total amount of energy of course remaining unaltered. The downfall of the energy accordingly consists in the breaking of its quanta into smaller units. And when once the fall and breakage have taken place, it is as impossible to reconstitute the original large quanta as it was to put Humpty-Dumpty back on his wall.

Although this is the main part of the downward path, it is not the whole of it. Thermodynamics teaches that all the different forms of energy have different degrees of "availability" and that the downward path is always from higher to lower availability.

And now we may return to the question with which we started the present chapter: "what is it that keeps the varied life of the universe going?" Our original answer "energy" is seen to be incomplete. Energy is no doubt essential, but the really complete answer is that it is the transformation of energy from a more available to a less available form; it is the running downhill of energy. To argue that the total energy of the universe cannot diminish, and therefore the universe must go on for ever, is like arguing that as a clock-weight cannot diminish, the clock-hand must go round and round for ever.

BEGINNINGS AND ENDINGS

The Final End of the Universe

Energy cannot run downhill for ever, and, like the clock-weight, it must touch bottom at last. And so the universe cannot go on for ever; sooner or later the time must come when its last erg of energy has reached the lowest rung on the ladder of descending availability, and at this moment the active life of the universe must cease. The energy is still there, but it has lost all capacity for change; it is as little able to work the universe as the water in a flat pond is able to turn a water-wheel. We are left with a dead, although possibly a warm, universe—a "heat-death."

Such is the teaching of modern thermodynamics. There is no reason for doubting or challenging it, and indeed it is so fully confirmed by the whole of our terrestrial experience, that it is difficult to see at what point it could be open to attack. It disposes at once of any possibility of a cyclic universe in which the events we see are as the pouring of river water into the sea, while events we do not see restore this water back to the river. The water of the river can go round and round in this way, just because it is not the whole of the universe; something extraneous to the river-cycle keeps it continually in motion—namely, the heat of the sun. But the universe as a whole cannot so go round and round. Short of postulating continuous action from outside the universe, whatever this may mean, the energy of the universe must continually lose availability; a universe in which the energy had no further availability to lose would be dead already. Change can occur only in the one direction, which leads to the heat-death. With universes as with mortals, the only possible life is progress to the grave.

Even the flow of the river to the sea, which we selected as an obvious instance of true cyclic motion, is seen to illustrate this, as soon as all the relevant factors are taken into account. As the river pours seaward over its falls and cascades, the tumbling of its waters generates heat, which ultimately passes off into space in the form of heat radiation. But the energy which keeps the river pouring along comes ultimately from the sun in the form mainly of light; shut off the sun's radiation and the river will soon stop flowing. The river flows only by continually transforming light-energy into heat-energy, and as soon as the cooling sun ceases to supply energy of sufficiently high availability the flow must cease.

The same general principles may be applied to the astronomical universe. There is no question as to the way in which energy runs down here. It is first liberated in the hot interior of a star in the form of quanta of extremely short wave-length and excessively high energy. As this radiant energy struggles out to the star's surface, it continually adjusts itself, through repeated absorption and re-emission, to the temperature of that part of the star through which it is passing. As longer wave-lengths are associated with lower temperatures (p. 132), the wave-length of the radiation is continually lengthened; a few energetic quanta are being transformed into numerous feeble quanta. Once these are free in space, they travel onward unchanged until they meet dust particles, stray atoms, free electrons, or some other form of interstellar matter. Except in the highly improbable event of this matter being at a higher temperature than the surfaces of the stars, these encounters still further increase the wave-length of the radiation, and the final result of innumerable encounters is radiation of very great wave-length. The quanta have increased enormously in numbers, but have paid for their increase by

BEGINNINGS AND ENDINGS

a corresponding decrease in individual strength. In all probability, the original very energetic quanta had their source in the annihilation of protons and electrons, so that the main process of the universe consists in the energy of exceedingly high availability which is bottled up in electrons and protons being transformed into heat energy at the lowest level of availability.

Many, giving rein to their fancy, have speculated that this low-level heat energy may in due course reform itself into new electrons and protons. As the existing universe dissolves away into radiation, their imagination sees new heavens and a new earth coming into being out of the ashes of the old. But science can give no support to such fancies. Perhaps it is as well; it is hard to see what advantage could accrue from an eternal reiteration of the same theme, or even from endless variations of it.

The final state of the universe will, then, be attained when every atom which is capable of annihilation has been annihilated, and its energy transformed into heat-energy wandering for ever round space, and when all the weight of any kind whatever which is capable of being transformed into radiation has been so transformed.

We have mentioned Hubble's estimate that matter is distributed in space at an average rate of 1.5×10^{-31} grammes per cubic centimetre. The annihilation of a gramme of matter liberates 9×10^{20} ergs of energy, so that the annihilation of 1.5×10^{-31} grammes of matter liberates 1.35×10^{-10} ergs of energy. It follows that the total annihilation of all the substance of the existing universe would only fill space with energy at the rate of 1.35×10^{-10} ergs per cubic centimetre. This amount of energy is only enough to raise the temperature of space from absolute zero to a temperature far below that of liquid air; it would only raise the temperature of the earth's surface by a 6000th part of a degree Centigrade. The reason why the effect of annihilating a whole universe is so extraordinarily slight is of course that space is so extraordinarily empty of matter; trying to warm space by annihilating all the matter in it is like trying to warm a room by burning a speck of dust here and a speck of dust there. As compared with any amount of radiation that is ever likely to be poured into it, the capacity of space is that of a bottomless pit. Indeed, so far as scientific observation goes, it is entirely possible that the radiation of thousands of dead universes may even now be wandering round space without our suspecting it.

Such is the final end of things to which, so far as present-day science can see, the material universe must inevitably come in some far-off age, unless the course of nature is changed in the meantime. Let us now try to peer back towards the beginnings of things.

The Beginnings of the Universe

As we go forwards in time, material weight continually changes into radiation. Conversely, as we go backwards in time, the total material weight of the universe must continually increase. We have seen how the present weights of the stars are incompatible with their having existed for more than some 5 or 10 million million years, and that they would need approximately the whole of this enormous period to acquire certain signs of age which their present arrangement and motions reveal.

We have seen that the break-up of the huge extra-galactic nebulae must result in the birth of stars, and have found that the most consistent account of the origin of the galactic system of stars is provided by the supposition that the whole system originated out of the break-up of a single huge nebula some 5 to 10 million million years ago.

Let us pause for a moment to compare this with an alternative hypothesis, which some astronomers have favoured, that stars are being created all the time. On this hypothesis we picture the stars as passing in an endless steady stream from creation to extinction, just as men pass in an endless steady stream from their cradles to their graves, a new generation always coming into being to step into the place vacated by the old. On this view Plaskett's star, with about a hundred times the weight of the sun, must be a recent creation, while Kruger 60, with only a fraction of the sun's weight, would be very, very old—perhaps 100 million million years older than Plaskett's star.

At present direct observation cannot definitely decide between the two conflicting hypotheses, but it rather frowns upon the "steady stream" view of the stars. In a steady population the number of people in any assigned condition is exactly proportional to the time taken to pass through that condition. Suppose for instance that human beings possess infant teeth for a quarter as long as they possess adult teeth. If examination of the teeth of a population shewed that four times as many had adult teeth as infant teeth, this would create a *prima facie* expectation that we were dealing with a steady population. If, on the contrary, 100 times as many people were found with adult teeth as with infant teeth, we should know we were not dealing with a steady population. If other evidence pointed to the population all being of approximately the same age, we should be inclined to accept this and regard the 1 per cent. of cases of infant teeth as cases of arrested development.

We do not judge the ages of stars by their teeth but by their weights and luminosities. And the luminosities of the stars are not found to conform to the statistical laws which would prevail in a steady population of stars. There appear to be so many middle-aged stars and so few infants and veterans as to make the hypothesis of a steady continuous creation hardly tenable. Indeed there is rather distinct evidence of a special creation of stars at about the time our sun was born. This leads back again quite naturally to the view that the galactic system was born out of a spiral nebula whose main activity as a parent of stars occurred some 5 to 10 million million years ago.

Pre-Stellar Existence. On the whole it seems likely that we must assign ages of 5 to 10 million million years to most or all of the stars in the galactic system. This is as far as we can probe back into time with any reasonable plausibility. The atoms which now form the sun and stars must no doubt have had a previous existence as atoms of a nebula, but we cannot say for how long. The temperature at the centres of the spiral nebulae may be, and in all probability are, so high that atoms are stripped bare of electrons and so shielded from annihilation. We may in fact regard the gaseous centres of nebulae as a sort of "white-dwarfs" built on a colossal scale. This fits on to the fact that the nebulae generate very little energy for their weights and so shine very feebly.

We have seen that the weights of two extra-galactic nebulae can be estimated to a reasonable degree of accuracy. The great Andromeda nebula M 31 has the weight of 3500 million suns, its total luminosity being that of 660 million suns. The nebula N.G.C. 4594 has the weight of 2000 million suns, and the luminosity of 260 million suns. A simple calculation shews that the atoms in the Andromeda nebula have an average expectation of life of 80 million million years, while the corresponding figure in N.G.C. 4594 is 115 million million years. From these two instances, we may guess that the average life, before annihilation, of the atoms in such nebulae must be of the order of 100 million million years. It cannot be claimed that this calculation is either very convincing or very exact, but it supplies the only evidence at present available as to the probable length of life of matter in the nebular state. We can say that the stars have existed *as such* for from 5 to 10 million million years, and that their atoms may have previously existed in nebulae for at least a comparable, and possibly for a much longer, time.

Apart from detailed figures, however, it is clear that we cannot go backward in time for ever. Each step back in time involves an increase in the total weight of the matter of the universe, and, just as with individual stars, we cannot go so far back that this total weight becomes infinite. Indeed a limit may quite possibly be set by considerations which we have already mentioned. The complete annihilation of all the matter now in the universe would raise the temperature of the earth's surface by the six-thousandth part of a degree; the annihilation of a million times as much matter would raise it by 160 degrees. We cannot admit that as much radiation as this can be wandering about space. The earth's temperature is determined by the amount of radiation it receives from the sun; it adjusts its temperature so that it radiates away just as much energy as it receives. A small correction is required on account of the earth's own radio-activity, but this need not bother us. What would bother us, and would indeed upset the balance entirely, would be the radiation of a million dead universes if this were for ever streaming on to us out of space; in this event the earth's surface would have to rise to a temperature well above that of boiling water before it could restore the balance between the radiation it received and that emitted. In a word, the radiation of a million dead universes would boil our seas, rivers and ourselves.

The Creation of Matter. All this makes it clear that the present matter of the universe cannot have existed for ever: indeed we can probably assign an upper limit to its age of, say, some such round number as 200 million million years. And, wherever we fix it, our next step back in time leads us to contemplate a definite event, or series of events, or continuous process, of creation of matter at some time not infinitely remote. In some way matter which had not previously existed, came, or was brought, into being.

If we want a naturalistic interpretation of this creation of matter, we may imagine radiant energy of any wave-length less than 1.3×10^{-13} cms. being poured into empty space; this is energy of higher "availability" than any known in the present universe, and the running down of such energy might well create a universe similar to our own. The table on p. 135 shews that radiation of the wave-length just mentioned might conceivably crystallise into electrons and protons, and finally form atoms. If we want a concrete picture of such a creation, we may think of the finger of God agitating the ether.

We may avoid this sort of crude imagery by insisting on space, time, and matter being treated together and inseparably as a single system, so that it becomes meaningless to speak of space and time as existing at all before matter existed. Such a view is consonant not only with ancient metaphysical theories, but also with the modern theory of relativity (p. 74). The universe now becomes a finite picture whose dimensions are a certain amount of space and a certain amount of time; the protons and electrons are the streaks of paint which define the picture against its space-time background. Travelling as far back in time as we can, brings us not to the creation of the picture, but to its edge; the creation of the picture lies as much outside the picture as the artist is outside his canvas. On this view, discussing the creation of the universe in terms of time and space is like trying to discover the artist and the action of painting, by going to the edge of the picture. This brings us very near to those philosophical systems which regard the universe as a thought in the mind of its Creator, thereby reducing all discussion of material creation to futility.

Both these points of view are impregnable, but so also is that of the plain man who, recognising that it is impossible for the human mind to comprehend the full plan of the universe, decides that his own efforts shall stop this side of the creation of matter. This last point of view is perhaps the most justifiable of all from the purely philosophic standpoint. It is now a full quarter of a century since physical science, largely under the leadership of Poincaré, left off trying to explain phenomena and resigned itself merely to describing them in the simplest way possible. To take the simplest illustration, the Victorian scientist thought it necessary to "explain" light as a wave-motion in the mechanical ether which he was for ever trying to construct out of jellies and gyroscopes; the scientist of to-day, fortunately for his sanity, has given up the attempt and is well satisfied if he can obtain a mathematical formula which will predict what light will do under specified conditions. It does not matter much whether the formula admits of a mechanical explanation or not, or whether such an explanation corresponds to any thinkable ultimate reality. The formulae of modern science are judged mainly, if not entirely, by their capacity for describing the phenomena of nature with simplicity, accuracy, and completeness. For instance, the ether has dropped out of science, not because scientists as a whole have formed a reasoned judgment that no such thing exists, but because they find they can describe all the phenomena of nature quite perfectly without it. It merely cumbers the picture, so they leave it out. If at some future time they find they need it, they will put it back again.

This does not imply any lowering of the standards or ideals of science; it implies merely a growing conviction that the ultimate realities of the universe are at present quite beyond the reach of science, and may be—and probably are—for ever beyond the comprehension of the human mind. It is *à priori* probable that only the artist can understand the full significance of the picture he has painted, and that this will remain for ever impossible for a few specks of paint on the canvas. It is for this kind of reason that, when, as in Chapter II, we try to discuss the ultimate structure of the atom, we are driven to speak in terms of similes, metaphors, and parables. There is no need even to worry overmuch about apparent contradictions. The higher unity of ultimate reality must no doubt reconcile them all, although it remains to be seen whether this higher unity is within our comprehension or not. In the meantime a contradiction worries us about as much as an unexplained fact, but hardly more; it may or may not disappear in the progress of science.

If some such train of thought may be applied to our efforts to understand the most minute processes of the universe (and it is the common everyday train of thought of those who are working in this field), then it must surely be still more applicable to our efforts to understand the universe as a whole. Phenomena come to us disguised in their frameworks of time and space; they are messages in cypher of which we shall not understand the ultimate significance until we have discovered how to decode them out of their space-time wrappings. Whatever may be thought about our final ability to decode the difficult messages we have recently received about the ultimate structure of the minutest parts of matter, it seems natural that we should feel some apprehension with regard to those about the structure of the universe as a whole, and particularly those about its beginnings and endings. Often enough the message itself may help us to discover the code in which it reaches us—with sufficient skill we can often do this—but we are now speaking of problems as to when, by whom, and for what purpose, the code was devised. There is no reason why a code message should throw any light on this.

The astronomer must leave the problem at this stage. The message of astronomy is of obvious concern to philosophy, to religion and to humanity in general, but it is not the business of the astronomer to decode it. The observing astronomer watches and records the dots and dashes of the needle which delivers the message, the theoretical astronomer translates these into words—and according as they are found to form known consistent words or not, it is known whether he has done his job well or ill—but it is for others to try to understand and explain the ultimate decoded meaning of the words he writes down.

Life and the Universe

Abandoning our efforts to understand the universe as a whole, let us glance for a moment at the relation of life to the universe we know.

The old view that every point of light in the sky represented a possible home for life is quite foreign to modern astronomy. The stars themselves have surface-temperatures of anything from 1650° to 30,000° or more, and are of course at far higher temperatures inside. By far the greater part of the matter of the universe is at a temperature of millions of degrees, so that its molecules are broken up into atoms, and the atoms are broken up, partially at least, into their constituent parts. Now the very concept of life implies duration in time; there can be no life where atoms change their make-up millions of times a second and no pair of atoms can ever stay joined together. It also implies a certain mobility in space, and these two implications restrict life to the small range of physical conditions in which the liquid state is possible. Our survey of the universe has shewn how small this range is in comparison with that exhibited by the universe as a whole. It is not to be found in the stars, nor in the nebulae out of which the stars are born. We know of no type of astronomical body in which the conditions can be favourable to life except planets like our own revolving round a sun.

Now planets are very rare. They come into being as the result of the close approach of two stars, and stars are so sparsely scattered in space that it is an inconceivably rare event for one to pass near to a neighbour. Yet exact mathematical analysis shews that planets cannot be born except when two stars pass within about three diameters of one another. As we know how the stars are scattered in space, we can estimate fairly closely how often two stars will approach within this distance of one another. The calculation shews that even after a star has lived its life of millions of millions of years, the chance is still about a hundred thousand to one against its being a sun surrounded by planets.

Even so, if life is to obtain a footing, the planets must not be too hot or too cold. In the solar system, for instance, we cannot imagine life existing on Mercury or on Neptune; liquids boil on the former and freeze hard on the latter. These planets are unsuitable for life because they are too near to, or too far from, the sun. We can imagine other planets which are unsuitable because their substance itself generates energy at such a rate as to make them unsuitable for habitation. The inert atoms which form our earth seem to be the end products of a long series of atomic changes, a sort of final ash resulting from the combustion of the universe. We have seen how such atoms probably float to the top in every star, as being the lightest in weight, but it is by no means a foregone conclusion that all planets will consist of nothing but inert atoms, and so will cool down until life can obtain a footing on them. This has happened with our earth, but we do not know how many planets and planetary systems may be unsuited for life because it has not happened with them.

All this suggests that only an infinitesimally small corner of the universe can be in the least suited to form an abode of life. Primaeval matter must go on transforming itself into radiation for millions of millions of years to produce a minute quantity of the inert ash on which life can exist. Then by an almost incredible accident this ash, and nothing else, must be torn out of the sun which has produced it, and condense into a planet. Even then, this residue of ash must not be too hot or too cold, or life will be impossible.

Finally, after all these conditions are satisfied, will life come or will it not? We must probably discard the at one time widely accepted view that once life had come into the universe in any way whatsoever, it would rapidly spread from planet to planet and from one planetary system to another until the whole universe teemed with life; space now seems too cold, and planetary systems too far apart. Our terrestrial life must in all probability have originated on the earth itself. What we would like to know is whether it originated as the result of still another amazing accident or succession of coincidences, or whether it is the normal event for inanimate matter to produce life in due course, when the physical environment is suitable. We look to the biologist for the answer, which so far he has not been able to produce.

The astronomer might be able to give a partial answer if he could find evidence of life on Mars or some other planet, for we should then at least know that life had occurred more than once in the history of the universe, but so far no convincing evidence has been forthcoming. The supposed canals on Mars disappear when looked at through a really large telescope, and have not survived the test of being photographed. Seasonal changes necessarily occur on Mars as on the earth, and certain phenomena accompany these which many astronomers are inclined to ascribe to the growth and decline of vegetation, although they may represent nothing more than rains watering the desert. There is no definite evidence of life, and certainly no evidence of conscious life, on Mars—or indeed anywhere else in the universe.

It seems at first somewhat surprising that oxygen figures so largely in the earth's atmosphere, in view of its readiness to enter into chemical combination with other substances. We know, however, that vegetation is continually discharging oxygen into the atmosphere, and it has often been suggested that the oxygen of the earth's atmosphere may be mainly or entirely of vegetable origin. If so, the presence or absence of oxygen in the atmospheres of other planets should shew whether vegetation similar to that we have on earth exists on these planets or not.

Oxygen certainly exists in the Martian atmosphere, but its amount is small. Adams and St. John estimate that there cannot be more than 15 per cent. as much, per square mile, as on earth. On the other hand it is either completely absent, or of negligible amount, in the atmosphere of Venus. If any is present at all, St. John estimates that the amount above the clouds which cover the surface of Venus is less than 0.1 per cent. of the terrestrial amount. The evidence, for what it is worth, goes to suggest that Venus, the only planet in the solar system outside Mars and the earth on which life could possibly exist, possesses no vegetation and no oxygen for higher forms of life to breathe.

Apart from the certain knowledge that life exists on earth, we have no definite knowledge whatever except that, at the best, life must be limited to a tiny fraction of the universe. Millions of millions of stars exist which support no life, which have never done so and never will do so. Of the rare planetary systems in the sky, many must be entirely lifeless, and in others life, if it exists at all, is probably limited to a few planets. The three centuries which have elapsed since Giordano Bruno suffered martyrdom for believing in the plurality of worlds have changed our conception of the universe almost beyond description, but they have not brought us appreciably nearer to understanding the relation of life to the universe. We can still only guess as to the meaning of this life which, to all appearances, is so rare. Is it the final climax towards which the whole creation moves, for which the millions of millions of years of transformation of matter in uninhabited stars and nebulae, and of the waste of radiation in desert space, have been only an incredibly extravagant preparation? Or is it a mere accidental and possibly quite unimportant by-product of natural processes, which have some other and more stupendous end in view? Or, to glance at a still more modest line of thought, must we regard it as something of the nature of a disease, which affects matter in its old age when it has lost the high temperature and capacity for generating high-frequency radiation with which younger and more vigorous matter would at once destroy life? Or, throwing humility aside, shall we venture to imagine that it is the only reality, which creates, instead of being created by, the colossal masses of the stars and nebulae and the almost inconceivably long vistas of astronomical time?

Again it is not for the astronomer to select between these alternative guesses; his task is done when he has delivered the message of astronomy. Perhaps it is over-rash for him even to formulate the questions this message suggests.

The Earth and Its Future Prospects

Let us leave these rather abstract regions of thought and come down to earth. We feel the solid earth under our feet, and the rays of the sun overhead. Somehow, but we know not how or why, life also is here; we ourselves are part of it. And it is natural to enquire what astronomy has to say as to its future prospects.

The central facts which dominate the whole situation are that we are dependent on the light and heat of the sun, and that these cannot remain for ever as they now are. So far as we can at present see, solar conditions can hardly have changed much since the earth was born; the earth's 2000 million years form so small a fraction of the sun's whole life that we can almost suppose the sun to have stood still throughout it. This of itself suggests that, in so far as astronomical factors are concerned, life may look to a tenancy of the earth of far longer duration than the total past age of the earth.

The earth, which started life as a hot mass of gas, has gradually cooled, until it has now about touched bottom, and has almost no heat beyond that which it receives from the sun. This just about balances the amount it radiates away into space, so that it would stay at its present temperature for ever if external conditions did not change, and any changes in its condition will be forced on it by changes occurring outside.

These external changes may be of many kinds. The sun's loss of weight causes the earth to recede from it at the rate of about a yard a century, so that after a million million years, the earth will be 10 per cent. further away from the source of its light and life than now. Consequently even if the sun then radiated as much light and heat as now, the earth would receive 20 per cent. less of this radiation, and its mean temperature would be some 15 degrees of Centigrade or so lower than at present. But after a million million years the sun will not radiate as much light and heat as now; it will have lost some 6 per cent. of its present weight through radiation, and, judging from other stars, this loss will probably reduce its energy-generating capacity by about 20 per cent. This will reduce the earth's temperature by about another 15 degrees, so that after a million million years the inevitable course of events will have reduced the earth's temperature by about 30 degrees Centigrade.

It would be rash to attempt to predict how such a fall of temperature may affect terrestrial life, and human life in particular. Given sufficient time, life has such an enormous capacity for adapting itself to its environment that it seems possible that, even with a temperature 30 degrees Centigrade lower than now, life may still exist on earth a million million years hence. If so, I am glad that my life has not fallen in this far distant future. Mountains and seas, which provide some of the keenest pleasures of our present life, will exist only as traditions handed down from a remote and almost incredible past. The denudation of a million million years will have reduced the mountains almost to plains, while seas and rivers will be frozen packs of solid ice. We may well imagine that man will have infinitely more knowledge than now, but he will no longer know the thrill of pleasure of the pioneer who opens up new realms of knowledge. Disease, and perhaps death, will have been conquered, and life will doubtless be safer and incomparably better-ordered than now. It will seem incredible that a time could have existed when men risked, and lost, their lives in traversing unexplored country, in climbing hitherto unclimbed peaks, in fighting wild beasts for the fun of it. Life will be more of a routine and less of an adventure than now; it will also be more purposeless when the human race knows that within a measurable space of time it must face extinction, and the eternal destruction of all its hopes, endeavours, and achievements.

Without laying too much stress on these visionary concepts of life a million million years hence, we may nevertheless think of this as the period in round numbers after which the inevitable wastage of the sun's weight is likely to drive life off the earth. Venus, with a mean temperature some sixty degrees higher than the earth, is probably rather too hot for life at present. But after a million million years, the temperature of Venus will have fallen by forty degrees, and what the earth is now, Venus may perhaps be somewhere between one and two million million years hence. Whether life will then inhabit Venus we cannot know, and it would be futile to guess, but there is at least a chance that as the earth fails, Venus may step into its place. Possibly Venus may be followed by Mercury in due course, but the present evidence is that Mercury is devoid of atmosphere, in which case it is hard to imagine it as a home for life at all resembling that which now inhabits the earth.

So far we have considered only the normal course of events; a variety of accidents may bring the human race to an end long before a million million years have elapsed. To mention only possible astronomical occurrences, the sun may run into another star, any asteroid may hit any other asteroid and, as a result, be so deflected from its path as to strike the earth, any of the stars in space may wander into the solar system and, in so doing, upset all the planetary orbits to such an extent that the earth becomes impossible as an abode of life. It is difficult to estimate the likelihood of any of these events happening but they all seem very improbable, and the first and last highly so. Let us disregard them all.

A danger remains which cannot be so lightly dismissed. Let us first state it in technical language. The sun is a main-sequence star, and is moreover very near to the left-hand edge of the main-sequence in the Russell diagram (p. 261). Beyond this edge is a region of the diagram which is completely untenanted by stars. We have supposed this region to be untenanted by stars because the stellar configurations it represents would be unstable. Stars pass through it rapidly until they find a stable configuration, and so end up in a region which can be permanently tenanted by stars. Now the next stable configurations beyond this region are those of the white-dwarfs, and as these are less massive as a class than the main-sequence stars, the general trend of stellar evolution appears to be from main-sequence star to white-dwarf. On this view the white-dwarfs must have previously been main-sequence stars which wandered across the left-hand edge of the band of stable configurations and then fell through the unstable region until they resumed stability as white-dwarfs.

The danger lies in the fact that the sun is already perilously near to the left-hand edge of the main-sequence. According to Redman's determinations, which are probably by far the most reliable at present available, the main-sequence belt of stable configurations for stars of the same spectral type as the sun (G 0) extends roughly between stellar absolute magnitudes, 4.88 and 3.54, the former marking the dangerous left-hand edge. The sun's present absolute magnitude is estimated as 4.85. Thus if the sun were to become 0.03 magnitudes fainter, this representing a reduction of only 3 per cent. in its luminosity, it would arrive exactly at the edge of the main-sequence, and would proceed to contract precipitately to the white-dwarf state. In so doing, its light and heat would diminish to such an extent that life would be banished from the earth. The known white-dwarf star which it would most closely resemble is the companion of Sirius, and this emits only a four-hundredth part as much light and heat as the sun.

To put the same thing in non-technical language, the sun is in, or is not far from, a precarious state in which stars are liable to begin to shrink and in so doing to reduce their radiation to a tiny fraction of that at present emitted by the sun. The shrinkage of the sun to this state would transform our oceans into ice and our atmosphere into liquid air; it seems impossible that terrestrial life could survive. The vast museum of the sky must almost certainly contain examples of shrunken suns of this type with planets like our earth revolving round them. Whether these planets carry on them the frozen remains of a life which was once as active as our present life on earth we can hardly even surmise.

This may be thought to open up a startling prospect for the earth, but we can take courage for several reasons. In the first place a 3 per cent. decrease in the sun's luminosity can hardly occur in less than about 150,000 million years. This in itself is not too bad, but the prospect becomes enormously more hopeful when we reflect that the evolution of the stars, including the sun, takes place in a direction almost parallel to the edge of the main-sequence. The sun is not heading for the precipice, so much as skirting along its edge. Whether it is approaching the edge, and is ultimately destined to fall over, we do not know, but it is in any case unlikely to reach the edge within the next million million years.

Finally, the sun's distance from the edge of the main-sequence cannot be estimated with anything like the degree of accuracy assumed in the foregoing calculations. The figure of 0.03 appeared as the difference of two much larger numbers, and although both of these can be estimated with fair accuracy, neither can be estimated with sufficient accuracy to justify us in treating their small difference of 0.03 as exact. The most we can say is that the sun is quite fairly near to the dangerous edge, but that any appreciable motion towards this edge is a matter of millions of millions of years. On the whole, while it has to be admitted that accidents may happen, there seems to be no reason for modifying our round number estimate of a million million years as the probable expectation, in the light of what astronomical knowledge we at present possess, of the future life of the human race on earth.

This is some five hundred times the past age of the earth, and over three million times the period through which humanity has so far existed on earth. Let us try to see these times in their proper proportion by the help of yet another simple model. Take a postage-stamp, and stick it on to a penny. Now climb Cleopatra's needle and lay the penny flat, postage-stamp uppermost, on top of the obelisk. The height of the whole structure may be taken to represent the time that has elapsed since the earth was born. On this scale, the thickness of the penny and postage-stamp together represents the time that man has lived on earth. The thickness of the postage-stamp represents the time he has been civilised, the thickness of the penny representing the time he lived in an uncivilised state. Now stick another postage-stamp on top of the first to represent the next 5000 years of civilisation, and keep sticking on postage-stamps until you have a pile as high as Mont Blanc. Even now the pile forms an inadequate representation of the length of the future which, so far as astronomy can see, probably stretches before civilised humanity. The first postage-stamp was the past of civilisation; the column higher than Mont Blanc is its future. Or, to look at it in another way, the first postage-stamp represents what man has already achieved; the pile which outtops Mont Blanc represents what he may achieve, if his future achievement is proportional to his time on earth.

Looked at in terms of space, the message of astronomy is at best one of melancholy grandeur and oppressive vastness. Looked at in terms of time, it becomes one of almost endless possibility and hope. As denizens of the universe we may be living near its end rather than its beginning; for it seems likely that most of the universe had melted into radiation before we appeared on the scene. But as inhabitants of the earth, we are living at the very beginning of time. We have come into being in the fresh glory of the dawn, and a day of almost unthinkable length stretches before us with unimaginable opportunities for accomplishment. Our descendants of far-off ages, looking down this long vista of time from the other end, will see our present age as the misty morning of the world's history; our contemporaries of to-day will appear as dim heroic figures who fought their way through jungles of ignorance, error and superstition to discover truth, to learn how to harness the forces of nature, and to make a world worthy for mankind to live in. We are still too much engulfed in the greyness of the morning mists to be able to imagine, however vaguely, how this world of ours will appear to those who will come after us and see it in the full light of day. But by what light we have, we seem to discern that the main message of astronomy is one of hope to the race and of responsibility to the individual—of responsibility because we are drawing plans and laying foundations for a longer future than we can well imagine.

INDEX

α Aquilae, *see* Altair
α Canis Majoris, *see* Sirius
α Canis Minoris, *see* Procyon
α Centauri, distance, 32, 35
   luminosity and weight, 47
   system of, 39, 248, 263, 264
α Herculis, 244, 256
α Lyrae, 32
α Orionis, *see* Betelgeux
α-particles, 105, 109, 110, 112,
   Plate XIII (p. 106)
α-rays, 104
α Scorpii, *see* Antares
Absolute temperature, 96
Absorption lines and spectrum, 48,
   49, 118
Actino-uranium, 145, 146
Adams, J. C., 18, 19
Adams, W. S., 59, 323
Age of earth, 13, 142 ff.
   of stars, 146 ff.
   of sun, 171, 174
   of universe, 316
Altair (α Aquilae), 31, 32, 271
Andromeda, Great Nebula (*M* 31)
   in, Plate IV (p. 30), 29, 65,
   67, 195, 198, 314
   age, 315
   distance, 66
   rotation, 196
   size, 30
   weight, 66, 198, 204
Andromedid meteors, 236
Angstrom unit, defined, 241
Angular momentum, conservation
   of, 195, 207, 220, 221
   of solar system, 221
Annihilation of matter, 175, 180,
   181, 281, 311, 312
Antares (α Scorpii), 244, 256, 257,
   261
   internal constitution of, 257, 274

Aquarid meteors, 236
Aristarchus of Samos, 2, 13
Aristotle, 2, 5
Arrangement of solar system,
   227 ff.
   Bode's law, 19, 20, 237
   Copernican, 3, 5, 6, 26, 31
   Ptolemaic, 2, 4, 6, 8, 30, 31
Asteroids, 17, 215, 227, 233, 236,
   237, 327
   Ceres, 17
   distance of, 19, 20, 237
   Eros, 33
   origin of, 233, 236
Aston, F. W., 111, 145, 175
Atmosphere, of earth, 184, 322
   of moon, 186
   of planets, 186, 322, 323, 327
   of sun, 185
Atomic nuclei, 101, 110
   disintegration of, 104, 110
   size and weight of, 102, 103
   structure of, 112, 130, 131, 138
Atomic numbers and weights, 103,
   271
   in stellar matter, 288, 294
Atomic theory, 86, 97
Atoms, 97, 297
   structure of, 100
   synthesis of, 138, 139, 174
Availability of Energy, 306, 308

β Aurigae, 52, 55
β-particles, 106, 107, 109, 137,
   Plate XIII (p. 106)
β-rays, 104, 137
B.D. 6° 1309, *see* Plaskett's star
Bacon, Roger, 1
Becquerel, H., 104
Bentley, R., 187
Berenice's hair, 37
Bessel, F. W., 32

INDEX

Betelgeux (α Orionis), 244, 256, 261
Biela's comet, 236
Binary systems, 37, 38, 50, 51, 52
  birth of, 206
  eclipsing, 51, 52, 53, 55
  loss of weight in, 178, 217
  orbits of, 43, 157, 158
  origins of, 158, 206, 268
  relative weights of components in, 178
  subdivision in, 217, 218
  weights of, 46, 47, 52
Birth of binary systems, 206
  of nebulae, 193
  of planets, 220, 225
  of satellites, 220, 227
  of stars, 199
Blackett, P. M. S., 110, 111
Blue stars, 59
Bode's law, 19, 20, 237
Bohr, Niels, 117, 121, 122, 124, 287
Bowen, J. S., 136
Boys, C. V., 42
Brownian movements, 150, 154
Bruno, Giordano, 3, 4, 323
Buffon, G. L. L., 223
Bumstead, H. A., 107
Burton, E. F., 134

γ-rays, 104, 107, 109, 132, 134, 135, 137
Cameron, G. H., 136
Candle-power of stars, 47, 53, 169, 177; see Luminosity
Capella, 261
Cassini, D., 22
Cavity-radiation, 115, 240
Cepheid variables, 53, 54, 55, 56, 57, 58, 65, 199
  mechanism of, 212
Ceres (asteroid), 17
Chaos, primaeval, density of, 192
  evolution from, 192, 195, 218, 219
Clusters, globular, 60, 61, 62, 203, 205, 206
  moving, 38, 163, 205, 206
Coal, combustion of, 166, 180
Coal-sack, 29

Comets, 233, 236
  origin of, 225
Condensation-chamber of C. T. R. Wilson, 106, 110
Condensations in a gas, 188, 189, 190, 193, 198, 225
Configurations of rotating masses, 210
Conservation, of angular momentum, 195, 206, 220
  of energy, 93, 178, 306, 307
  of matter, 178
Constellations, 35, 37
Cooke, H. L., 134
Copernican astronomy, 3, 5, 6, 26, 31
Copernicus, 3
Creation, of matter, 316
  of universe, 73, 317
Crookes, Sir W., 100
Cycles and epicycles (Ptolemaic), 2, 3, 4
Cyclic processes in nature, 305, 308, 309, 310, 311

δ Cephei, 53, 212
δ-rays, 107
Darwin, Sir George, 213
de Sitter, cosmology of, 73 ff., 164
Democritus, 87
Density, of matter in space, 71, 311, 312
  of nebulae, 198
  of stars, 250-252, 257, 258, 266
Diameters of stars, 245, 248-252, 256, 257, 261, 263, 264
Diffraction grating, 108
Dimensions, of earth, 33
  of galactic system, 62
  of solar system, 19, 33
  of universe, 71, 77, 80, 81, 82
Discontinuity, in physics, 116 ff.
  of matter, 87
Displacements of spectral lines, (de Sitter), 75
  (Doppler), 49
  (Einstein), 73, 246, 303
Distances, of globular clusters, 60
  of nebulae, 66, 67
  of stars, 30, 31, 33, 34, 35, 58, 59, 60, 61

[334]

INDEX

Doppler effect, 49
Dwarf and Giant stars, 259
Dwarfs, white, 264, 267, 268, 275, 299, 300, 302, 304, 314

Earth, age of, 13, 142, 146, 164
  as a planet, 16, 19, 216, 227, 325
  birth of, 225, 230
  dimensions of, 33
  future of, 214, 215, 234, 324, 329, 330
  -moon system, 33, 214, 215, 227, 229, 230, 235
  orbit of, 33, 34, 216, 325
Eccentricity of binary orbits, 161
  of ellipse (defined), 44
Eclipsing binaries, 51, 52, 53, 55
Eddington, A. S., 66, 169, 174, 177, 213, 271, 278
Einstein, A. (Cosmology), 72, 73, 74
  (Gravitation), 43, 70, 73, 151
  (Quantum theory), 117, 119, 120
  (Relativity), 71 ff., 176, 246, 303
Electricity, positive and negative, 100, 101
  attraction and repulsion, 100, 121
Electromagnetic energy, 113, 114
Electron, 100, 106, 112, 114
  orbits in atom, 101, 102, 120
Ellipse, defined, 44
  as atomic orbit, 121
  as gravitational orbit, 44, 216
Ellis, C. D., 131
Emden, R., 270, 271
Energy, 92, 113, 178, 179, 306 ff.
  availability of, 306, 308
  conservation of, 93, 178
  source of stellar, 174
  weight of, 113
Epicycles, 2, 3, 4
Equipartition of energy, in a gas, 146, 147
  stellar, 150, 151, 154, 160, 161
Eros, 33
Ether, existence of, 318

Evaporation, 88
Evening stars, 17
Evolution, from primaeval chaos, 192, 195, 218, 219
  stellar, 295 ff.
Extra-galactic nebulae, 27, 29, 65, 66
  densities of, 198
  distance of, 66, 67
  evolution of, 194 ff., and Plates XVI to XXI in order
  number of, 67
  rotation of, 195, 196
  size of, 30
  velocities of, 76, 77, 78
  weights of, 66, 204

Faraday, M., 115
Final end of universe, 309 ff.
Fission of stars, 159, 206, 211, 216, 218
Fizeau, H. L., 246
Fluorescence, 108, 307
Fraunhofer, J. von, 99
Free path of molecules, 91
Frequency of radiation, 108, 119
Full radiator, 116, 240, 246

Galactic latitude and longitude, 23, 25
Galactic nebulae, 28, Plates III (p. 29), VI (p. 36), VII (p. 42)
Galactic system of stars, 22, 25, 62, 63, 201
  number of, 63, 64, 204
  weight of, 64, 204
Galileo, 1, 3, 4, 5, 8, 16, 22, 23
Galle, J. G., 19
Gaseous masses in rotation, 196, 210
Gaseous state, nature of, 89
  equipartition of energy in, 146, 147
Geodesy, 32
Geology, 141, 143
Giant and Dwarf stars (defined), 259
Giant stars, internal constitution, 257, 273, 274, 291, 292, 299

INDEX

Globular clusters, 60, 61, 62, 203, 205, Plate IX (p. 60)
  origin of, 206
  shape of, 203
Goldsbrough, G. R., 235
Graham, T., 149
Gravitation, law of, 39, 43, 72, 151, 152
  cause of, 70
Gravitational instability, 187, 200, 225, 232
Great Bear (cluster), 163, 205, 206
  (constellation), 30, 37
Great nebulae, see Extra-galactic nebulae

H.D. 1337, see Pearce's star
Halley, E., 142
Halley's comet, 233, 236
Halm, *J., 153
Hayford, J. F., 33
Heat, nature of, 94
  effect of, on electrical structures, 132, 135
  transport of, in a star, 289
Heat-death, 309, 310
Helium atom, 103, 112, 174, 175, 287
Helmholtz, H., 167, 269
Henderson, T., 32
Herschel, Sir John, 22, 24, 201
Herschel, Sir William, 18, 22, 23, 24, 25, 26, 27, 30, 62, 65, 201
Hertzsprung, E., 56, 212, 259
Hesperus, 17
Highly penetrating radiation, 134, 135, 181, 281, 282
Hinks, A. R., 62
Holmes, A., 143
Hubble, E., 65, 66, 71, 77, 192, 194, 196, 197, 200
Huyghens, C., 22
Hydrogen atom, 103, 112, 123, 127, 138
  annihilation of, 138, 181-182
  electron orbits in, 121, 122, 123, 182

Instability, dynamical, 191
  gravitational, 187, 200, 225, 232

Interferometer, 245, 246, 257
Interstellar matter, 28-29, 60
Interstellar space, temperature of, 96
Inverse square law (electric), 121
  (gravitational), 42, 43, 44
  (luminosity), 24, 31
Invisible radiation, 109, 243, 244
Irregular nebulae, 196
Irregular variables, 53, 55
Island universes, 27, 30, 65
Isotopes, 111

Jeffreys, H., 146, 214, 215, 226
Jupiter, 4, 16, 17, 19, 196
  birth of, 227, 228, 229, 230
  rotation of, 158, 214
  satellites of, 4, 22, 186, 228, 229, 230, 235
  size of, 18
  temperature of, 20
  weight of, 43

K-ring of electrons in atom, 126, 130, 272, 273, 274, 275, 276, 285, 287, 288
Kant, I., 166, 219
Kapteyn, J. C., 26, 62, 201, 203
Kelvin, Lord, 167, 269
Kepler, J., 5, 31, 44
Klein, O., 137, 139
Kolhörster, W., 134
Kramers, H. A., 291, 292
Kruger, 46, 261, 263, 264, 299
  system of, 252, 263, 264

L-ring of electrons in atom, 126, 127, 130, 273, 275, 285
Lalande 21185, 35, 249
Lane's law, 269
Laplace, P. S., 219, 222, 223
  nebular hypothesis of, 219, 220, 223
Lead, radioactive, 145
Leavitt, Miss, 54
Lenard, P., 100
Leonid shooting-stars, 236
Leslie, Sir John, 149
Leucippus, 87
Leverrier, V. J., 18, 19

[336]

INDEX

Life in solar system, 320
  in universe, 319, 321, 323, 329
  on Earth, 12, 322, 324, 325, 326
  on Mars, 322
  on Venus, 323, 327
Light, nature and composition of, 35, 49, 107
  speed of, 35, 108
  wave-length of, 108, 130
Light-year, 36
Lippershey, 1
Liquid masses in rotation, 208 ff.
Liquid stars, 211, 279 ff., 283, 286, 287, 292, 293, 304
Lives of stars, see Stars
Lockyer, Sir N., 296
Long-period variables, 54, 55, 59, 255, 256, 301
Lucifer, 17
Lucretius, 87
Luminosity of stars, 46, 47, 169, 253, 254, 261, 263, 264

M-ring of electrons in atom, 126, 127, 273, 275, 276, 285
McLennan, J. C., 134
Magellanic clouds, 54, 78, 79, 200
  and Plate XXI (p. 204)
Main-sequence stars, 263, 275, 278
Major planets, 16, 17
  atmospheres on, 187
Man's life on earth: past, 12, 16, future, 324 ff.
Marius, 29
Mars, 18, 19, 35
  atmosphere of, 186, 322, 323
  birth of, 227, 229, 230
  life on, 322
  rotation of, 214
  satellites of, 227, 229, 235
  temperature of, 21
Matter, annihilation of, 175, 180, 181, 298
  conservation of, 178
Maxwell, J. C., 115, 148
Mayer, R., 166
Mercury, 4, 5, 16, 17, 19, 34
  absence of satellites, 229
  atmosphere of, 186, 327
  birth of, 227, 230
  life on, 327

Mercury—*Continued*
  orbit of, 43
  rotation of, 22, 215
Michelson, A. A., 246
Milky way, 3, 23, 24, 25, 26, 27, 29
Millikan, R. A., 134, 139
Minor planets, 16
Molecule, 88
  collisions of, 91, 92, 96, 185
  equipartition of energy, 146, 147
  speed of, 90, 185
  size of, 89
Moon, 4, 41
  atmosphere of, 186
  birth of, 227, 229
  distance of, 33, 41, 42
  eclipse of, 33
  future of, 214, 215, 235
  rotation of, 214
Morning star, 17
Moving clusters, 38, 163, 205
  shape of, 205, 206
Munich 15040, 34, 249

Nebulae, classification of, 28
Nebulae, extra-galactic, 27, 29, 30, 65, 193 ff.
  distance and size of, 29, 30, 66, 67
  rotation of, 195
  weights of, 67, 194, 198, 204
  M 31, see Andromeda, Great Nebula in
  M 33, 66, 200 and Plate XX (p. 201)
  N.G.C. 4594, 66, 67, 194, 198 and Plate XV (p. 194)
Nebulae, galactic, 29 and Plates III (p. 29), VI (p. 36), VII (p. 42)
Nebulae, planetary, 28, 257, 303, 314 and Plate II (p. 28)
Nebulae, spiral, 29, 224
Nebular hypothesis of Laplace, 219, 220, 223
Neptune, 16, 18, 19
  birth of, 226, 229, 230
  rotation of, 214
  satellite of, 227, 229, 230
  temperature of, 20

[337]

INDEX

New stars (novae), 59
Newton, Sir Isaac (cosmogony), 187, 193
  (distances of stars), 31, 32
  (law of gravitation, 41, 42, 43, 44, 45, 151, 187
  (light and optics), 48, 98
Nicholas of Cusa, 3
Nishina, Y., 137, 139
Novae, 59
Nucleus, atomic, *see* Atomic

o Ceti, 55, 256, 257, 264
o_2 Eridani, 253, 261, 264
O-type stars, 303
Obscuring matter in galactic system, 23, 28, 29
Opacity of stellar matter, 290
Orbit, of binary systems, 44, 45, 157 ff., 213 ff.
  of earth, 33, 34, 216
  of moon, 33, 41, 215, 235
  of planets, 42, 43, 225
  of stars, 201, 202, 205, 206
Oresme, 2
Orion, constellation of, 31, 37, 163

Parallactic motion, defined, 32
Parallactic measures of stellar distances, 32, 58, 60
Parallax, spectroscopic, 60, 285
Pearce's star (H.D. 1337), 46, 52, 299
Period-luminosity law (Cepheid-variables), 57-59
Perrin, J., 150, 174
Perseid shooting-stars, 236
Phases of moon and inner planets, 4, 5
Philolaus, 2
Phosphoros, 17
Photo-electric action, 120
Photography in astronomy, 36
Piazzi, G., 17
Planck, M., 116, 118, 119, 240
Planetary motions, 34, 43, 44, 73
Planetary nebulae, 28, 257, 303 and Plate II (p. 28)
Planetary orbits, 42, 43, 225, 231
Planets, arrangement of, 16, 19, 227

Planets—*Continued*
  birth of (Laplace), 220, 222, 223
  birth of (tidal theory), 225, 226, 227, 230
  in the universe, 320, 328, 329
  motions and orbits of, 34, 35, 42, 43, 72, 225, 231
Plaskett's star (B.D. 6° 1309), 46, 52, 253, 255, 256
Pleiades, 30, 37, 38, 163, 205
Plummer, H. C., 213
Poincaré, H., 269, 317
Poincaré's theorem, 269, 275
Primaeval chaos, evolution from, 192, 195, 218, 219
Pressure in a gas, 91
  in a star, 289
  of radiation, 113
  of radiation in a star, 274, 289
Procyon, 46, 251, 261, 263, 264
Proton, 112, 114
Proxima Centauri, 34, 38, 248, 249, 261
Ptolemy, 2, 3, 4, 5, 6, 8, 30
Pythagoras, 2, 3, 5, 13

Quantum, defined, 119
Quantum-theory, 115 ff., 129 ff., 281, 310, 311
Quotations, Arnold, Matthew, 5
  Bede, 7
  Bruno, Giordano, 3
  Cornford, Frances, 11
  Cusa, Nicholas of, 3
  Galileo, 5
  Newton, Sir Isaac, 187
  Sackville-West, Victoria, 7
  Shakespeare, 183

Radiation, 107 ff., 129 ff., 239 ff., 290
  distribution by wave-length, 240
  highly penetrating, 134, 135, 136, 181, 281, 282
  mechanical effects of, 113, 129, 272, 289
  of sun, 52, 165, 241
  pressure of, 113; in a star, 274, 288

[338]

INDEX

Radiation—*Continued*
  visible and invisible, 243, 244
  wave-length of, 115, 240, 243
  weight of, 113, 114, 175, 176
Radio-activity, 104 ff., 144, 281
Rainbow, 48, 98
Rayleigh, Lord, 88
Redman, H. O., 260, 328
Relativity, theory of, 69 ff., 176, 246, 303, 317
Ring nebula in Lyra, 28, 258, and Plate II (p. 28)
Rings of electrons, in atom, 126, 130, 272, 273, 274 ff., 285, 287, 288
Roche, E., 234
Roche's limit, 232, 234, 235
Rosse, Lord, 29
Rotation, of astronomical bodies, 158, 205
  of galaxy, 203
  of nebulae, 195
  of stars, 206
Rotating masses, of gas, 196, 210
  of liquid, 208, 210
Rotating nebulae, 195
Rotating systems of stars, 158, 203
Royds, T., 105
Russell, H. N., 177, 218, 259, 276, 282, 297
Russell diagram, 260, 261, 262, 266, 267, 275, 279, 283, 285, 286, 295, 296, 297, 327
Rutherford, Sir E., 101, 104, 105, 110, 134, 139, 145

S Doradus, 172, 254
St. John, C. E., 323
Salinity of oceans, 142
Sampson, R. A., 18, 290
Satellites, atmospheres on, 186
  birth of (Laplace), 220
  birth of (tidal theory), 227, 228, 232
  discovery of, 22
Saturn, 16, 18, 19, 31, 43
  birth of, 227, 228, 229, 230
  rings of, 215, 220, 233, 235
  rotation of, 214
  satellites of, 22, 186, 227, 228 230, 233, 234, 235

Saturn—*Continued*
  temperature of, 20
Schrödinger, E., 287
Schwarzschild, K., 290
Seares, F. H., 26, 63, 154, 265, 266
Sedimentation, 142
Shapley, H., 26, 56, 60, 62, 63, 203, 205, 213
Shooting-stars, nature of, 166, 233, 234
  origin of, 225, 233, 234
  swarms of, 236
Sirius, 34, 46, 244, 261, 263, 264, 299
  system of, 249
Sirius B, 46, 246, 264, 275
Soddy, F., 104
Solar spectrum, 48, 98, 118, 241
Solar system, arrangement of,
  Bode's law, 19
  arrangement of (Copernican), 3, 5, 6, 26, 31
  arrangement of (Ptolemaic), 2, 4, 6, 8, 31
  atmospheres in, 184 ff., 322, 323, 326
  orbits in, 42, 43, 213 ff., 225
  origin of, 219 ff.
Sound, 90, 188
Spectra, stellar, 48, 118, 123 and Plate VIII (p. 48). *See also* Displacements
Spectral types, 48
  relation to stellar weights, 154
  relation to surface temperatures, 242
Spectroscope, 48
Spectroscopic binaries, 50
  orbits of, 161, 162
  origin of, 159, 211, 212
Spectroscopic parallaxes, 60, 285
Spectroscopic velocities, 48
Spectroscopy, 98
Spectrum, 48
  of sun, 48, 99, 118, 241
Spherical nebulae, 196, 207, 208
Spiral nebulae, 29, 224
  arms of, 224
Stability of stellar structures, 278, 279, 282, 283

INDEX

Stars, arrangement of, 23, 25, 26
 ages of, 78, 146, 157, 163, 164, 176
 birth of, 199
 density of, 248 ff., 257, 266, 275
 diameters of, 245 ff., 261, 262, 301
 distances of, 30, 35
 evolution of, 295 ff.
 internal constitution of, 268 ff.
 internal temperatures of, 270
 liquid, 211, 279, 280, 281, 289
 luminosity of, 47, 169, 170, 254
 motions of, 34, 48, 63, 154, 201, 210
 number of, 63, 83
 spectra of, 48, 118, 123 and Plate VIII (p. 48). *See also* Displacements
 stability of, 278, 279, 282, 285, 286
 surface temperatures of, 239, 241, 242, 247 ff., 254, 255, 256
 variable, 53 ff., 213
 weights of, 43, 45, 46, 47, 154, 169, 253, 264 ff.
Star-streaming, 209, 211
Statistical methods in astronomy, 201, 202
Stellar, *see* Stars
Struve, W., 32
Struve, Otto, 212
Sun, age of, 171, 174
 candle-power of, 53
 distance from earth, 33, 216, 325
 future of, 327, 328
 internal constitution of, 270, 273
 internal temperature of, 270, 271, 272
 loss of weight of, 168, 216, 325
 past history of, 170
 position in galactic system, 25, 63
 radiation of, 53, 165, 179, 180
 rotation of, 158, 195, 231
 surface-temperature of, 241
Surface temperatures of planets, 20 ff., 325, 326
 of stars, 239, 242, 248 ff., 254, 255, 256
 of sun, 241

Taylor's comet, 236
Telescope, 1
 Galileo's, 2, 13
 Herschels', 24
 Lord Rosse's, 29
 100-inch, 63, 68
 200-inch, 68
Temperature, scale of, 96
Temperature-radiation, 131, 240
Thermodynamics, 306 ff.
 first and second laws of, 306, 307
Thomson, Sir J. J., 100
Tidal friction, 213 ff.
Tidal theory of solar system, 224 ff.
Tides in binary stars, 213, 215
Titan (satellite of Saturn), 22, 186
Transit of Venus, 33
Transport of energy in a star, 289

Ultra-violet radiation, 109, 134, 135
Universe, age of, 314, 315
 beginning of, 73, 312 ff.
 final end of, 309 ff.
 size of, 77, 79, 81
 structure of, 68, 81
Uranium atom, 104
 disintegration of, 109, 113, 114, 144
Uranus, 16, 18, 19
 birth of, 226, 229, 230
 rotation of, 214
 satellites of, 22, 227, 229, 230
 temperature of, 20

V Puppis, 261, 299
van Maanen, A., 258
van Maanen's star, 252, 258, 261, 275, 304
van Rhijn, P. J., 25
Variable stars, 53 ff., 211, 256, 301
Vega (α Lyrae), 32
Velocities of stars, 34, 153 ff.
Venus, 4, 5, 16, 17, 19, 34
 absence of satellites, 230
 atmosphere on, 187, 323
 birth of, 226, 230
 life on, 323, 327

INDEX

Venus—*Continued*
  phases of, 4, 5
  rotation of, 213
  temperature of, 21
  transit of, 34
Virgo, nebulae N.G.C. 4594 in, 66, 67, 194, 198 and Plate XV (p. 194)
Visual binaries, orbits of, 160, 161
  origin of, 159, 268

Wave-length of radiation, 108, 115

Wave-mechanics, 128, 287
White-dwarf stars, 264, 267, 276, 299, 300, 304, 328
Wilson, C. T. R., 106, 110
Wireless transmission, 109
Wolf 359, 34, 249, 254

X-rays, 109, 130, 135, 289, 293

Zero (absolute) of temperature, 96
Zodiacal light, 226